FINITE MARKOV CHAIN AND FUZZY LOGIC ASSESSMENT MODELS

Emerging Research and Opportunities

Michael Gr. Voskoglou, Ph.D.

Professor Emeritus of Mathematical Sciences
School of Technological Applications
Graduate Technological Educational Institute (T. E. I.) of Western Greece

M. Voskoglou: Finite Markov Chain and Fuzzy Logic Assessment Models

Dedicated to my beloved grandchildren Michalis, Alexandra, Sygouros and Angeliki, my hope for the future

PREFACE

INTRODUCTION AND ACKNOWLEDGMENTS

There used to be a tradition in science and engineering of turning to probability theory when one is faced with a problem in which uncertainty plays an important role. This transition was justified when there were no alternative tools for dealing with the uncertainty. Today this is no longer the rule. ***Fuzzy logic***, which is based on fuzzy sets theory introduced by Zadeh (1965), due to its property of characterizing the ambiguous cases with multiple values provides a rich and meaningful addition to the traditional bi-valued logic that opens the door to construction of mathematical solutions of computational problems which are stated in a natural language. In contrast, standard probability theory does not have this capability, a fact which is one of its principal limitations. The applications which may be generated from or adapted to fuzzy logic are wide-ranging and provide the opportunity for modelling under conditions which are inherently imprecisely defined, despite the concerns of classical logicians. Many systems may be modelled, simulated and even replicated with the help of fuzzy logic, not the least of which is human reasoning itself.

On the other hand, ***Markov chains***, which are a smart combination of Linear Algebra and Probability Theory, offer ideal conditions for the study and mathematical modelling of a certain kind of phenomena depending upon random variables. The basic concepts of Markov chains were introduced by Markov in 1907 for application on coding literary texts. Since then the Markov chain theory was developed by a number of leading mathematicians such as Kolmogorov, Feller etc, but only from the 60's the importance of this theory to the natural, social and most of the other applied sciences has been realized (Suppes & Atkinson, 1960, Kemeny & Snell, 1963, 1976, Bartholomew, 1973, etc.).

The material presented in this book is an extension of my Lecture Notes (Voskoglou, 2016) of the Course No. 161021K03 of the ***Global Initiative on Academic Network (GIAN)*** Program of the Government of India, which took place on October, 2016, at the Mathematics Department of the ***National Institute of Technology (NIT) of Durgapur, India***. This material

M. Voskoglou: Finite Markov Chain and Fuzzy Logic Assessment Models

is a product of a research work of the author and its collaborators starting during the early 1990's and continuing until today. The outcomes of this research include more than 170 articles published in reputed scientific journals and proceedings of International Conferences of about 30 countries of the five continents around the world, with very many citations by other researchers.

The book consists of the Preface and eight Chapters. The Preface, apart from the introduction to the subject and the acknowledgments, includes a list of symbols and abbreviations in alphabetical order and the Table of Contents. The first Chapter introduces the reader to the art of Mathematical Modelling. The next two Chapters concern applications of finite Markov Chains to Management and to learning contexts. The following three Chapters include applications of Fuzzy Logic to learning contexts and not only. Finally, Chapter 8 involves a recapitulating application, our general conclusions and hints for future research. A list of references is provided at the end of the Preface and of each of the book's eight Chapters.

The primary objectives of the book are the following:

i) Exposing the reader to the fundamentals of finite Markov Chains, of Uncertainty theory in fuzzy environment and of Fuzzy Sets and Numbers. As the emphasis is given to applications rather, the theoretical proofs proceed only up to the point which is indispensable for the better understanding of the related contents of the book.

ii) Building confidence and capability to the readers for applying tools and techniques of Markov Chains and Fuzzy Logic and for mapping the organizational activities and problems in terms of the above mathematical topics.

iii) Providing exposure to practical problems and their solutions through case studies that can be modelled by utilizing principles of Ergodic/Absorbing Markov Chains and of Fuzzy Sets theory.

iv) Enhancing the reader's capability to identify, control and adapt related problems to learning contexts, to Management, to assessment of player performance and to other relative topics.

Apart from the secondary education knowledge and critical thinking, the only mathematical background needed for the better understanding of the book is some fundamentals of Probability Theory and a skilful knowledge of the basic operations between matrices.

The book is suitable for:

1 Mathematicians, computer science specialists, engineers and other applied scientists, as well as for researchers from manufacturing, service and government organizations.

2 Students at all levels (B.Tech./ M.Sc./ M.Tech./ Ph.D.) and Faculties from academic and technical institutions.

3 From those possessing the least required mathematical background and interested to study applications of mathematics to real world problems and in particular to human cognitive contexts.

The author wishes to express his sincere thanks to the ***Indian Government*** for the the GIAN grant and to his close collaborator ***Dr. Anita Pal***, Assistant Professor at the Department of Mathematics of NIT of Durgapur, for her kind invitation to visit and lecture at her Institute and for her great contribution to the organization of the corresponding GIAN course. He wishes also to thank the participants of the above course (Faculty members, undergraduate, postgraduate and research students and industry researchers) for their valuable remarks and comments, which helped him enormously for the final formulation of the present book's contents. Finally, but not less, the author wishes to thank ***Prof. Dr. Igor Ya. Subbotin***, State University, Los Angeles, USA, for the excellent research collaboration that they have all the years since 2011. A significant part of the material presented in this book is based on his own ideas and efforts. Thanks are also due to ***Prof. Dr. Yannis Theodorou***, Graduate T. E. I. of Central Greece , author in Greek language of a very interesting book on Fuzzy Logic and its Applications to Technology (Theodorou, 2010), for his useful comments on Fuzzy Numbers and to ***Dr. Steve Perdikaris***, for his collaboration on Makov Chains during the early 1990s (Voskoglou & Perdikaris, 1991, 1993).

Hoping that the book at hand will be useful for those to whom it is addressed, the author will accept with pleasure any comments and suggestions aiming to its further improvement.

June, 2017 - Patras, Greece

M. Voskoglou: Finite Markov Chain and Fuzzy Logic Assessment Models

Michael Gr. Voskoglou, Ph.D.
Professor Emeritus of Mathematical Sciences
Graduate Technological Educational Institute (T.E.I.) of Western Greece
School of Technological Applications, 26334 Patras- Greece
Tel.-Fax: 00302610328631, Mobile: 00306978600391
E-mail: <voskoglou@teiwest.gr> ; <mvosk@hol.gr>
Home page : http://eclass.teipat.gr/eclass/courses/523102

TABLE OF CONTENTS

LIST OF SYMBOLS AND ABBREVIATIONS

(In Alphabetical Order)

$A = [p_{ij}]$: Transition matrix of a Markov Chain

$A^* = \begin{bmatrix} I & | & O \\ - & | & - \\ R & | & Q \end{bmatrix}$**:** Partition of the canonical (or standard) form A* of the transition *n* X *n* matrix A of an absorbing Markov Chain with *k* absorbing states, where *I* denotes the unitary *k*X *k* matrix, *O* is a zero matrix, *R* is the *(n – k)* X *k* transition matrix from the non absorbing to the absorbing states and *Q* is the *(n – k)* X *(n – k)* transition matrix between the non absorbing states.

A^{-1}: The inverse matrix of the square matrix A

adj(A): The adjoin matrix of the square matrix A

A(t): A fuzzy variable representing a fuzzy set for each value of t = 1, 2,…,k.

AMC: Absorbing Markov Chain

AR: Analogical Reasoning

CBR: Case - Based Reasoning

COG: Centre of Gravity

CrT: Critical Thinking

CT: Computational (Algorithmic) Thinking

D(A): Determinant of the square matrix A

DM: Decision – Making

EPM: Expert Performance Model (for Problem – Solving)

EMC: Ergodic Markov Chain

FL: Fuzzy Logic

FN: Fuzzy Number

M. Voskoglou: Finite Markov Chain and Fuzzy Logic Assessment Models

FS: Fuzzy Set

***f(s)*:** Pseudo – frequency of the profile *s*

GPA: Grade Point Average

GRFAM: Generalized Rectangular Fuzzy Assessment Model

H: Probabilistic Uncertainty (Shannon's Entropy)

$m_A(x)$: Membership degree of *x* in the fuzzy set *A*

maximin of payoffs: Maximization of the minimal payoffs

MC: Markov Chain

M=[mij]; Mean first passage matrix of an EMC

MM: Mathematical Modelling

MPSF: Multidimensional Problem –Solving Framework

$m_s = m_R(s)$: Membership degree of the profile *s* defined with respect to the fuzzy relation *R*.

m. u. : Metric unit

$N = [n_{ij}]$: Fundamental matrix of an AMC

***N(r)*:** Non – specificity (or imprecision) on the ordered possibility distribution *r*

$P = [p_1\ p_2\ \ldots\ldots\ p_n]$: Limiting probability vector of a finite EMC with n states.

p_{ij}: Transition probability from state S_i to state S_j

$P_k = [p_1^{(k)}\ p_2^{(k)}\ \ldots\ldots\ p_n^{(k)}]$: Probability vector of the k-th phase of a finite MC with n states.

p_s: Probability of the profile *s*

PS: Problem - Solving

RFAM: Rectangular Fuzzy Assessment Model

r_s: Possibility of the profile *s*

***ST(r)*:** Strife (or discord) on the ordered possibility distribution *r*

TFAM: Triangular Fuzzy Assessment Model

TFN: Triangular Fuzzy Number

TpFAM: Trapezoidal` Fuzzy Assessment Model

M. Voskoglou: Finite Markov Chain and Fuzzy Logic Assessment Models

TpFN: Trapezoidal Fuzzy Number

***T(r)*:** Total possibilistic uncertainty on the ordered possibility distribution *r*

REFERENCES

Bartholomew, D.J. (1973), *Stochastic Models for Social Processes*, J. Wiley and Sons, London.

Kemeny, J. G. & Snell, J. L. (1963), *Mathematical Models in the Social Sciences*, Ginn and Company, New York, USA.

Kemeny, J. G. & Snell J. L. (1976), *Finite Markov Chains*, Springer - Verlag, New York, USA.

Suppes, P. & Atkinson, R. C. (1960), *Markov Learning Models for Multiperson Interactions*, Stanford University Press, Stanford-California, USA.

Theodorou, J., *Introduction to Fuzzy Logic*, Tzolas Publications, Thessaloniki, Greece, 2010 (in Greek language).

Voskoglou, M. Gr. & Perdikaris, S. C. (1991), A Markov chain model in problem- solving, *International Journal of Mathematical Education in Science and. Technology*, 22, 909-914.

Voskoglou, M. Gr. & Perdikaris, S. C. (1993), Measuring problem solving skills, *International Journal of Mathematical Education in Science and. Technology*,, 24, 443-447.

Voskoglou, M. Gr. (2016), *Finite Markov Chain and Fuzzy Models in Management and Education,* GIAN Program, Course No. 16102K03/2015-16, National Institute of Technology, Durgapur, India

Zadeh, L. (1965), Fuzzy Sets, *Information and Control,* 8, 338-353.

PART I

FINITE MARKOV CHAINS

CHAPTER 1
Mathematical Modelling

ABSTRACT

In this introductory Chapter the basic principles of the art of modelling are presented. The kinds of models in use are described with emphasis to mathematical models. The process of mathematical modelling is described and the structure of a mathematical model is studied. The notion of probability is also introduced and the stochastic models - a special kind of mathematical models involving random variables - are discussed. As an example, the Chapter closes with the development of a simple model for the process of learning a subject matter in classroom, based on the notion of conditional probability. The background needed for a better understanding of the Chapter is presented at the end in the form of Endnotes, following the list of References.

THE ART OF MODELLING

The notion of a ***system*** has a very broad context. Roughly speaking, it can be defined as a set of interacting components forming an integrated whole. Examples of systems include the physical systems (the Earth, our solar system, the whole universe, etc), social systems (our society, religions, countries and organizations, scientific communities, etc), economic systems (companies, industries, etc), biological systems like human or animal organizations, abstract systems (mathematical, philosophical, etc), artificial systems designed by the humans (buildings, transportation means, etc) and many others.

The systems' modelling is a basic principle in engineering, in natural and in social sciences. When we face a problem concerning a system's operation (e.g. maximizing the productivity of an organization, minimizing the functional costs of a company, etc) a ***model*** is required to describe and represent the system's multiple views. The model is a simplified

representation of the basic characteristics of the real system including only its entities and features under concern. The construction of a model usually involves a preliminary deep abstracting process on identifying the system's dominant variables and the relationships governing them (***system's analysis***). The resulting structure of this action is known as the ***assumed real system***. The model, being a further abstraction of the assumed real system, identifies and simplifies the relationships among these variables in a form amenable to analysis. The above process is sketched in Figure 1 (Taha, 1967).

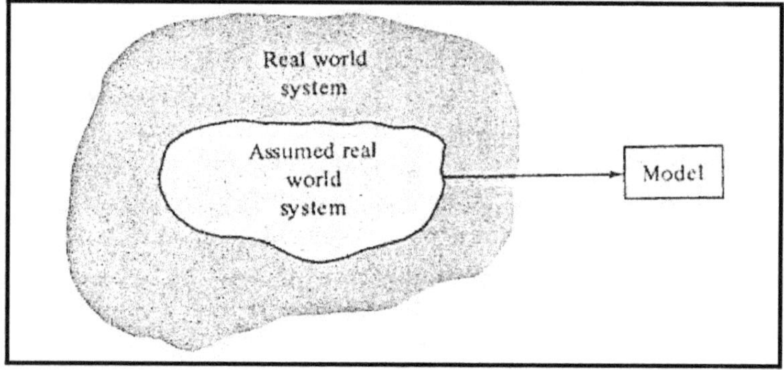

Figure 1: A graphical representation of the modelling process

KINDS OF MODELS

There are several types of models in use according to the form of the corresponding problem. The representation of a system's operation through the use of a ***mathematical*** (or ***symbolic***) ***model*** is achieved by a set of mathematical expressions (equalities, inequalities, etc) and functions properly related to each other. The solutions provided by a mathematical model are more general and accurate than those provided by the other types of models.

However, in cases where a system's operation is too complicated to be described in mathematical terms (e.g. biological systems), or the corresponding mathematical relations are too difficult to deal with in providing the problem's solution, a ***simulation model*** can be used. Such a model consists of a sequence of logical orders properly designed to mimic the real system's behaviour and usually operating with the help of computers. On the contrary, for the representation of relatively simple

systems, ***figurative models*** are usually used, e.g. maps, bas-relief representations, etc. ***analogical models*** are also constructed for representing the quantitative relation between two system's variables, e.g. graphs, diagrams, etc. Finally, ***heuristic models*** are used when the invention of new strategies is requested for the improvement of already existing solutions.

MATHEMATICAL MODELS

Davies and Hersh (1981) state that "the usefulness of a mathematical model is exactly its success to foresee, or (and) to imitate accurately the behaviour of the real world". Until the middle of 1970's ***Mathematical Modelling (MM)*** was mainly a tool in hands of scientists and engineers for solving the real world problems related to their disciplines (physics, industry, constructions, economics, etc). One of the first who described the process of MM in such a way that it could be used for teaching purposes was Pollak (1979). He represented the interaction between mathematics and the real world with the scheme shown in Figure 2, which is known as the ***circle of MM***.

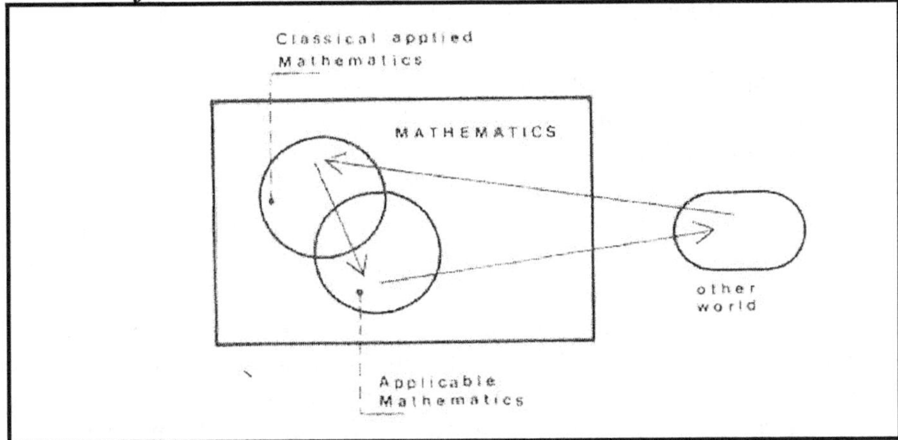

Figure 2: The circle of MM

According to the Pollak's scheme, in the "universe" of mathematics ***classical applied mathematics*** and ***applicable mathematics*** are two intersected, but not equal to each other sets. In fact, the first set contains all mathematical topics having real world applications. Nevertheless, as the history of mathematics shows, there are also branches of theoretical mathematics with no practical applications for the moment that could find such applications in future [1]. This kind of mathematics is characterized by Pollak as applicable mathematics However, the most important feature of

Pollak's scheme is the direction of the arrows, representing a looping between the other (real) world -including all the other sciences and the human activities of everyday life - and the "universe" of mathematics: Starting from a problem of the real world we transfer to the other part of the scheme, where we use or develop suitable mathematics for its solution. Then we return to the real world interpreting and testing on the real situation the mathematical results obtained. If these results are not giving a satisfactory solution to the given problem, then we repeat the same circle again one or more times.

From the time that Pollak presented this scheme in ICME-3 (Karlsruhe, 1976) until nowadays much effort has been placed to analyze in detail the process of MM. A comprehensive account of the different models used for the description of the MM process can be found in Haines & Crouch (2010) including the present author's Markov chain model (Voskoglou, 1994, 1995, 2007).

As a result of all these research efforts it is more or less acceptable nowadays that the process of MM involves the following steps:

- ***Analysis*** of the problem
- ***Mathematization*** (construction of the model)
- ***Solution*** of the model
- ***Validation*** (control) of the model
- ***Implementation*** of the final mathematical results to the real system.

More details about the steps of MM will be presented in the second section of Chapter 3. Note that, some authors consider further steps in the MM process; e.g. some of them divide mathematization to the steps of the ***formulation*** of the problem in a way that it will be ready for mathematical treatment and of the construction of the model, others divide the validation to the steps of ***interpretation*** and ***evaluation*** of the model, others add the stage of ***refining*** the model, etc. (Haines & Crouch, 2010). However, all these minor variations do not change the general idea that we nowadays have about the circle of MM.

In contrast to the other kind of models, mathematical models have a concrete structure involving the following elements:

- The ***decision variables*** representing the unknown quantities, the values of which are determined by the solution of the corresponding problem and the ***parameters*** (if any), which are variables whose values depend on external conditions that affect the system's behavior (e.g. prices of the required materials, of the fuel, etc.).

- The ***constraints***, being equations/inequalities between the decision variables that, following the existing in the system conditions, must take their ***feasible*** values only; for example if x represents a product's quantity, then x must take non negative values.

- The ***objective function***, which is a function of the decision variables and the parameters, the best value of which determines the corresponding problem's solution.

PROBABILITY - STOCHASTIC MODELS

In probability theory a ***random experiment*** is understood to be any action that can be repeated under the same conditions, in which all its possible outcomes are known, but the particular outcome of each repetition cannot be predicted.

Consider a random experiment with a *finite number* of mutually exclusive outcomes, which are *equally like* to happen and let *A* be an event associated with the possible outcomes of the experiment. Then the ***probability*** *P(A)* of the event *A* is defined to be the fraction of the outcomes in which *A* occurs. More explicitly, $P(A) = \frac{N(A)}{N}$, where *N* is the total number of outcomes of the experiment and *N(A)* is the number of outcomes leading to the occurrence of the event *A*.

For example, in case of throwing a single unbiased die we have $N = 6$. Let *A* be the event of getting an even number of spots. Then $N(A) = 3$ and $P(A) = \frac{3}{6}$, or 50%.

The above definition, usually referred as the ***mathematical definition of probability***, can be extended to cover experiments with outcomes which are not equally like to happen, provided that the corresponding experiment can be repeated under the same conditions any number of times. In this case, let n be the total number of experiments in the whole series of trials and let $n(A)$ be the number of experiments in which A occurs. Then the ratio $\frac{n(A)}{n}$ is called the ***relative frequency*** of the event A and the probability of A is defined by $P(A) = \lim_{N \to \infty} \frac{N(A)}{N}$ (***statistical definition of probability***).

The mathematical definition of probability, although does not require, in contrast to the statistical one, the performance of the corresponding experiment, it cannot be applied when the outcomes of the experiment are not equally like to happen (equiprobable) or/and when these outcomes are infinitely many. Obviously, in case where both definitions can be applied they coincide to each other [2]. For general facts on probability theory we refer to the book of Rozanov (1972).

Next, a variable is called a ***random variable***, if it takes values from a probability distribution. Any process involving at least one random variable is called a ***stochastic process***. In particular, a mathematical model in which at least one of its decision variables is a random variable is called a ***stochastic model***. Examples of stochastic models include the classical probability models, the Markov chain models (see the next two Chapters), which are based to a successful combination of Probability Theory and Linear Algebra, etc.

As an example, we shall develop in the next section a probability model for the process of learning a subject matter in the classroom.

A PROBABILITY MODEL FOR THE PROCESS OF LEARNING

Learning can be commonly defined as the activity of gaining knowledge or skill. The ability to learn is possessed by humans, animals, plants (Korban, 2015) and computers (Samuel, 1959). Learning does not happen all at once, but it builds upon and is shaped by previous knowledge. To that end, learning may be viewed as a process, rather than a collection of factual and procedural knowledge. In psychology and education learning refers to a process that brings together cognitive, emotional and environmental influences and experiences for acquiring, enhancing or making changes in one's knowledge, skills, values and world views (Ormod, 2012).

The process of learning is fundamental to the study of human cognitive action. But, while everyone knows empirically what learning is, the understanding of its nature has proved to be complicated. This happens because it is very difficult for someone to understand the way in which the human mind works, and therefore to describe the mechanisms of the acquisition of knowledge by the individual.

There are very many theories and models designed for the description of the mechanisms of learning. Volumes of research have been written about it and many attempts have been made by psychologists, cognitive scientists and educators to make learning accessible to all in various ways. There are three main philosophical frameworks under which learning theories fall: ***Behaviorism***, ***Cognitivism*** and ***Constructivism***. Behaviorism focuses only on the objectively observable aspects of learning; for behaviorists learning is the acquisition of new behavior through conditioning. Cognitive theories look beyond behavior to explain brain-based learning, while constructivism views learning as a process in which the learner actively constructs or builds new ideas and concepts.

Voss (1987) developed an argument that learning consists of successive problem – solving activities, in which the input information is represented of existing knowledge, with the solution occurring when the input is appropriately interpreted. The whole process involves the following steps:

- ***Representation*** of the stimulus input, which is relied upon the individual's ability to use contents of his (her) memory to find information, which will facilitate a solution development.
- ***Interpretation*** of the input data, through which the new knowledge is obtained.
- ***Generalization*** of the new knowledge to a variety of situations;
- ***Categorization*** of the generalized knowledge, so that the individual becomes able to relate the new information to his (her) already existing knowledge structures, which are referred as ***schemas***, or ***scripts***, or ***frames***.

The present author, based on Voss's framework for learning, developed a probability model for assessing the student progress during the learning of a subject matter in the classroom by considering the ***conditional***

probabilities [3] of suitably chosen events. An application of this method was presented in the first Mediterranean Conference on Mathematics Education in Cyprus (Voskoglou, 1997).

The experiment took place at the Graduate Technological Educational Institute (TEI) of Patras, Greece, when I was teaching to a group of 30 students of the School of Technological Applications (future engineers) the use of the derivative for the maximization and minimization of a function. During my 2 hours lecture I used the method of ***rediscovery*** (Polya, 1963) [4]. Thus, after a short introduction to the subject, I left my students to work alone on their papers. I was inspecting their works, and from time to time I was giving them the suitable instructions, or hints. After the basic theoretical conclusions and in order to check if they were able to generalize the new knowledge, I gave them some exercises to solve. At the final step some problems including applications to artificial constructions and economics were also given to them for solution (categorization of the new knowledge).

During the experiment I realized that four students didn't manage to understand satisfactorily the subject. Further five students, although it seemed that they had understood the basic theoretical ideas, they were unable to apply them for the solution of the given exercises and problems. The remaining 21 students solved the exercises, but 11 of them they didn't succeed to solve (or solved a small part only) of the given problems.

Consider now the following facts:

- A: A student interprets successfully the new information.
- B: A student solves the exercises (generalization).
- C: A student solves both the exercises and the problems (categorization).

Using the data of our observations and applying the mathematical definition of probability one finds that $P(A) = \frac{26}{30} \approx 0.8667$, $P(B) = P(A \cap B) =$

$= \frac{21}{30} = 0.7$, $P(C) = P(A \cap C) = P(B \cap C) = \frac{10}{30} \approx 0.3333$. Also one can calculate the conditional probabilities $P(B / A) = \frac{P(A \cap B)}{P(A)} = \frac{21}{26} \approx 0.8077$, $P(C / A) = \frac{P(C \cap A)}{P(A)} = \frac{10}{26} \approx 0.3846$ and $P(C / A) = \frac{P(C \cap B)}{P(B)} = \frac{10}{21} \approx 0.4762$.

Consequently, the probability for a student to interpret successfully *in the classroom* the new information is approximately equal to 86.67%, to generalize it is equal to 70% and to categorize the new knowledge is approximately equal to 33.33%. Further, the probability for a student, who has interpreted the new information, to be able to generalize the new knowledge to a variety of situations is approximately equal to 80.77 % and to be able to categorize it is approximately equal to 38.46%.. Finally, the probability for a student, who has generalized successfully the new knowledge, to be able to relate it to his (her) already existing schemas of knowledge (categorization) is approximately equal to 47.62 %.

One must note that he above probability values give an *approximate idea only* of the student progress during the learning process in the classroom. In fact, the above presented assessment method is based on the hypothesis that the solution of the exercises corresponds to the step of generalization and the solution of the problems to the step of categorization. However, this is a simplification of the real situation (assumed real system; Section 1.1), since the process of learning is very complicated depending on each individual's personal characteristics. On the other hand, the learning of a new subject matter by the student does not happen only in the classroom, but also afterwards, when doing other activities and even during sleeping!

REFERENCES

Davis, P. & Hersh, R. (1981), *The Mathematical Experience*, Birkauser Ed., Boston.

Haines, C. & Crouch, R. (2010), Remarks on a Modelling Cycle and Interpretation of Behaviours. In R.A. Lesh et al. (Eds.): *Modelling Students' Mathematical Modelling Competencies* (ICTMA 13), 145-154, Springer, USA.

Korban, R. (2015), Plant Learning and Memory, in: *Plant Sensing and Communication*, pp. 31-44, The University of Chicago Press, Chicago and London

Livio, M. (2009), *Is God a Mathematician?* Simon & Schuster, London

Ormod, J. (2012), *Human Learning*, 6th Edition, Pearson, Boston.

Pollak H. O. (1979), The interaction between Mathematics and other school subjects, *New Trends in Mathematics Teaching*, Volume IV, Paris: UNESKO.

Polya, G. (1945), *How to solve it,* Princeton Univ. Press, Princeton.

Polya, G. (1954), *Mathematics and Plausible Reasoning* (2 Volumes) , Princeton Univ. Press, Princeton.

Polya G. (1962/65), Mathematical Discovery (2 Volumes), J.Wilet & Sons, New York.

Rozanov, Y. A. (1972), *Probability Theory: A Concise Course* (Revised English Edition Translated and Edited by R. A. Silverman), Dover Publications, New York, USA.

Samuel, A. L. (1959), Some studies in machine learning using the game of checkers, *IBM Journal of Research and Development.*

Taha, H. A. (1967), *Operations Research - An Introduction*, 2nd Edition, Collier – Macmillan, New York – London.

Voskoglou, M. Gr. (1994), An application of Markov chains to the process of modelling, *International Journal of Mathematical Education in Science and. Technology*, 25, 475-480.

Voskoglou, M. Gr. (1995), Measuring Mathematical Model Building Abilities, *International Journal of Mathematical Education in Science and. Technology*, 26, 29-35.

M. Voskoglou: Finite Markov Chain and Fuzzy Logic Assessment Models

Voskoglou, M. Gr. (1997), Some Remarks on the Use of Rediscovery in the Teaching of Mathematics, *Proceedings of the 1st Mediterranean Conference on Mathematics Education,* 124-128, Nicosia, Cyprus.

Voskoglou, M. Gr. (2007), A stochastic model for the modelling process. In: *Mathematical Modelling: Education, Engineering and Economics* (ICTMA 12), Haines, Chr., Galbraith, P., Blum W. and Khan, S. (Eds), 149-157, Horwood Publishing, Chichester, England.

Voss, J. F. (1987), Learning and transfer in subject matter learning: A problem solving model, *International Journal of Educational Research,* 11, 607-622.

ENDNOTES

[1] The success of mathematics in the natural sciences appears in two forms, which were termed by Mario Livio (2009) as the "energetic" and the "pathetic" one respectively. In the former case scientists express the laws of nature mathematically by using relations and equations developed for this certain purpose. The effectiveness of mathematics in this case does not look so surprising, because the relative mathematical theories are designed to fit to the corresponding observations. On the contrary, the effectiveness of the pathetic form is really amazing. In this case completely abstract mathematical theories, developed without any intention to be applied in real life situations, are utilized in unsuspicious time for the construction of physical models! ***Knot Theory***, initiated from a false model for the description of the atom's structure, provides an amazing example of the interaction between the energetic and pathetic side of mathematics. In fact, the effort of mathematicians to understand the knots themselves, led finally to the conclusion that their theory was the key for understanding the basic mechanisms of the DNA! Another characteristic example is the use by Einstein of the ***Riemann's non Euclidean Geometry*** for developing the ***General Relativity Theory.*** This made Einstein to wonder: "How is it possible for mathematics, a derivative of the human mind independent from our experiences, to fit so eminently to the natural reality?"

[2] In 1930 the famous Russian mathematician Andrey Nikolaevich Kolmogorov introduced the following ***axiomatic definition of probability***: Let X be the set of all possible outcomes of a random experiment (***probability space***) and let S be the set of all possible events (subsets) of X. Then a ***modulus of probability on X*** is defined to be any function P: $S \to [0, 1]$, such that $P(X) = 1$ and $P(A \cup B) = P(A)+P(B)$, for all A, B in S

such that $A \cap B = \varnothing$. For example, in tossing a coin, there are two possible outcomes, namely "heads"= H and "tails"= T. Set {H}= *A* and {T}= *B*. Then, *X* = {H, T} and *S* = {*A*, *B*, *X*, $\varnothing$}. But $A \cap B = \varnothing$ and *A* $\cup$ *B* = *X*. Therefore, the above definition gives that *P(A)+P(B)* =*1*. Hence, each of the infinitely many non negative solutions of the equation x + y = 1 (e.g. x = $\frac{1}{3}$, y = $\frac{2}{3}$, or x = $\frac{\sqrt{2}}{2}$, y = 1- $\frac{\sqrt{2}}{2}$, etc.) defines a modulus of probability on *X*. In particular, the solution x = y = $\frac{1}{2}$ corresponds to the mathematical definition of probability, where the coin is supposed to be well-balanced. Although the axiomatic is a generalization of the mathematical and statistical definitions of probability not having their disadvantages, it does not offer a straight way of calculating probability, neither it insures its uniqueness.

[3] Let *A* be an event of the probability space *X* and let *B* be an event of the probability space *Y*, such that $A \cap B \neq \varnothing$ (Figure 3).

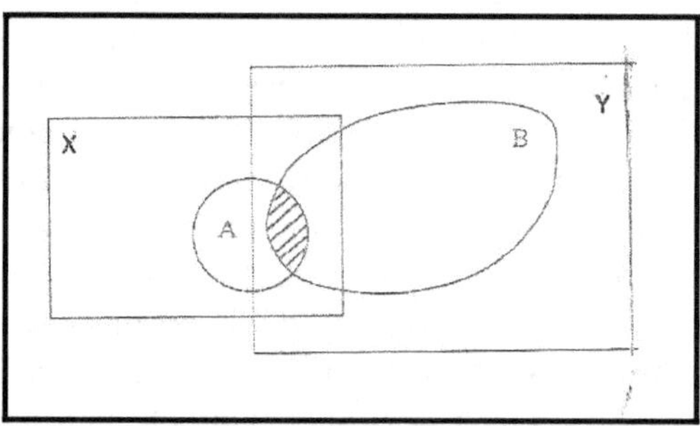

Figure 3: Conditional Probability

Denote by $A|B$ the event of *A* occurring under the condition that *B* is known to have occurred. The probability space of $A|B$ is obviously the set *B*. Assume further that the probability space Y is finite with equally like to happen (equiprobable) outcomes and let *N*, n_1 and n_2 be the numbers of elements of *Y*, $A \cap B$ and *B* respectively. Then, the mathematical definition of probability gives that

$$P(A|B) = \boxed{\frac{n_1}{n_2} = \frac{\frac{n_1}{N}}{\frac{n_2}{N}} = \frac{P(A \cap B)}{P(B)}} . (1)$$

The probability *P(A|B)* defined by relation (1) is called the ***conditional probability of A on the hypothesis B***. Definition (1) holds also for the case of statistical probabilities.

[4] According to Polya (1962/1965) "for an effective learning the learner discovers alone the biggest possible under the circumstances part of the new information". Polya's model for learning mathematics is based on his famous ***axioms of learning*** which include: ***(i) Active learning*** based on the solution by the learner of properly designed problems and on his/her effective discussion with the instructor in the form of the Socrates midwifery method. ***(ii)Best motivation***, which means that the instructor must find the proper way (e.g. an introductory problem, a historic event, an important connection with already existing knowledge, etc.) to create the proper "atmosphere" (learning situation), so that the learner will accept the new information with interest and pleasure. (iii) ***Consecutive phases***, which must be *exploration*, *formalization* and *assimilation* of the new mathematical knowledge.

Polya (1945, 1954) was also the first who introduced the ***heuristic strategies*** for problem-solving, i.e. general hints or suggestions that help solvers to understand first and then to solve a given mathematical problem. Such a heuristic, for example, states that, when you find difficult to solve a problem, try first to solve a similar simpler problem and then to adapt the solution found to solve the original problem (e.g. for solving an algebraic problem involving many variables try first to solve the corresponding problem in fewer variables, for solving a geometric problem in space try first to solve the corresponding problem in the level, etc.).

CHAPTER 2 Finite Ergodic and Absorbing Markov Chains

ABSTRACT

Markov chains offer ideal conditions for the study and mathematical modelling of a certain kind of situations depending on random variables. The main target of this Chapter is to make the reader able to apply the basic principles of the theory of finite Markov Chains for solving practical problems of the day to day life. Most of these problems can be solved by distinguishing between two types of finite Markov Chains, the Ergodic and the Absorbing ones. As the emphasis is given to the applications rather, the mathematical proofs of the relative propositions are omitted, unless if they are necessary for the better understanding of the corresponding subject matter.

BASIC DEFINITIONS

For the better understanding of this chapter the reader must be familiar with the basics of ***Probability Theory*** (Rozanov, 1972) and of ***Linear Algebra*** (Morris, 1978). The examples presented in this chapter were taken from Voskoglou (2016).

Roughly speaking, a ***Markov Chain (MC)*** is a stochastic process that moves in a sequence of steps (phases) through a set of states and has a "one-step memory", i.e. the probability of entering a certain state in a certain step, although in practice may not be completely independent of previous steps, depends at most on the state occupied in the previous step [1]. This is known as the ***Markov property***. When the set of its states is a finite

set, then we speak about a ***finite*** MC. For general facts on finite MC we refer to the classical on the subject book of Kemeny & Snell (1976).

The basic concepts of MC were introduced by A. Markov in 1907 on coding literary texts. Since then the MC theory was developed by a number of leading mathematicians, such as A. Kolmogorov, W. Feller etc. However, only from the 1960's the importance of this theory to the Natural, Social and most of the other Applied Sciences has been recognized (Suppes & Atkinson, 1960, Kemeny & Snell, 1963, Bartholomev, 1973, etc.). Most of the problems concerning applications of finite MC can be solved by distinguishing between two types of chains, the ***ergodic*** and the ***absorbing*** ones.

THE GENERAL MARKOV CHAIN MODEL

Let us consider a finite MC with *n* states, say $s_1, s_2, \ldots, s_n$, where *n* is a non negative integer, $n \geq 2$. Denote by p_{ij} the *transition probability* from state s_i to state s_j, $i, j = 1, 2, \ldots, n$; then the matrix $A = [p_{ij}]$ is called the ***transition matrix*** of the MC. Since the transition from a state to some other state (including itself) is the certain event, we have that $p_{i1} + p_{i2} + \ldots + p_{in} = 1$, for $i=1, 2, \ldots, n$.

The row-matrix $P_k = [p_1^{(k)}\ p_2^{(k)} \ldots\ p_n^{(k)}]$, known as the ***probability vector*** of the MC, gives the probabilities $p_i^{(k)}$ for the MC to be in state *i* at step *k* , for $i = 1, 2, \ldots, n$ and $k = 0, 1, 2, \ldots$ Obviously we have again that

$$p_1^{(k)} + p_2^{(k)} + \ldots + p_n^{(k)} = 1.$$

The following statement (Proposition 1) enables one to make ***short run forecasts*** for the evolution of various situations that can be represented by a finite MC. The proof of the Proposition, which is sketched below, serves to the understanding by the reader of the strict connection between MCs and Probability Theory.

1. ***Proposition:*** Under the previous notation we have that $P_{k+1} = P_k A$, for all non negative integers k.

Proof: We shall show first that $P_1 = P_0 A$ (1).

For this, consider the random variable $x = s_1, s_2, \ldots, s_n$, where s_i, $i = 1, 2, \ldots, n$ are the states of the MC and denote by $F_0, F_1, \ldots, F_k, \ldots$ the MC's consecutive phases. Consider the events:

- E: x takes the value s_1 at F_1, and
- E_i: x takes the value s_i at F_0, $i = 1, 2, \ldots, n$.

Then $E_i \cap E_j = \varnothing$, $i \neq j$, while E happens each time together with one, and only one, of the events $E_1, E_2, \ldots, E_n$. Therefore, by applying the ***total probability formula*** [2] we obtain that $P(E) = \sum_{i=1}^{n} P(Ei)P(E/E_i)$, where $P(E/E_i)$ denote the corresponding conditional probabilities.

But $P(E) = p_1^{(1)}$, $P(E_i) = p_i^{(0)}$ and $P(E/E_i) = p_{i1}$. Therefore, $p_1^{(1)} = p_1^{(0)}p_{11} + p_2^{(0)}p_{21} + \ldots + p_n^{(0)}p_{n1}$ and in the same way $p_i^{(1)} = p_1^{(0)}p_{1i} + p_2^{(0)}p_{2i} + \ldots + p_n^{(0)}p_{ni}$, $i = 1, 2, \ldots, n$ (2).

Writing the system of equations (2) in matrix form we obtain (1).

Further, replacing s_1 with s_k in the event E and working similarly, one can show in general that $P_{k+1} = P_k A$, for all non negative integers k.

2. ***Corollary:*** Under the above notation it is $P_n = P_0 A^n$, for all integers $n \geq 1$.

Proof: We apply induction on n. In fact, for $n = 1$ the statement $P_1 = P_0 A$ is true by Proposition 1. Assume further that it is true for $n = k$, i.e. that $P_k = P_0 A^k$. Then, by Proposition 1 again, $P_{k+1} = P_k A = (P_0 A^k) A = P_0 A^{k+1}$, which completes the proof.

The following simple example illustrates the above results:

3. ***Example:*** A company circulates for first time in the market a new product, say K.

The market's research shows that the consumers buy on average one such product per week, either K, or a competitive one. It is also expected that 70% of those who buy K they will prefer it again next week, while 20% of those who buy another competitive product they will turn to K next week. Find the market's share for K two weeks after its first circulation, provided that the market's conditions remain unchanged.

Solution: We form a MC having the following two states: s_1 = the consumer buys K, and s_2 = the consumer buys another competitive product. Then, the transition matrix of the MC is

$$A = [p_{ij}] = \begin{array}{c} \\ s_1 \\ s_2 \end{array} \begin{array}{c} \begin{array}{cc} s_1 & s_2 \end{array} \\ \begin{bmatrix} 0.7 & 0.3 \\ 0.2 & 0.8 \end{bmatrix} \end{array}$$

Further, since K circulates for first time in the market, we have that $P_0 = [0\ 1]$, therefore $P_2 = P_0 A^2 = [0.3\ 0.7]$ [3].

Thus the market's share for K two weeks after its first circulation will be 30%.

ERGODIC MARKOV CHAINS (EMCs)

A MC is said to be an ***Ergodic MC (EMC)***, if it is possible to go between any two states, not necessarily in one step. It is well known (Kemeny & Snell, 1976, Chapter 5) that, as the number of its steps tends to infinity (***long run***), an EMC tends to an ***equilibrium situation***, in which the probability vector P_k takes a constant price

$P = [p_1\ p_2\ \ldots.\ p_n]$, called the ***limiting probability vector*** of the EMC. Thus, as a direct consequence of Proposition 2.2.1, the equilibrium situation is characterized by the equality $P = PA$ (3), with $p_1 + p_2 + \ldots. + p_n = 1$. The entries of P express the probabilities of the EMC to be in each of its states in the long run.

The ***mean first passage matrix*** of an EMC is defined to be the n X n matrix M = $[m_{ij}]$, where m_{ij} is equal to the mean number of steps needed to go from state S_i to S_j for first time, I, j = 1, 2, …., n. It is well known then that $m_{ij} = \frac{p_i}{p_j}$, where Pi and Pj are the corresponding limiting probabilities (Kemeny & Snell, 1976, Chapter 5).

4. Example: Reconsider the statement in Example 3 and find:

(i) The market's share for the product K when the consumers' preferences are stabilized (long run).

(ii) The mean number of times that a costumer buys a competitive product between two successive purchases of K.

Solution: (i) The MC constructed in Example 2.2.3 is an EMC. Applying equation (3) for this EMC we get $[p_1\ p_2] = \begin{bmatrix} 0.7 & 0.3 \\ 0.2 & 0.8 \end{bmatrix} [p_1\ p_2]$, which gives that $p_1=0.7p_1+0.2p_2$ and $p_2=0.3p_1+0.8p_2$, or equivalently $0.3p_1 - 0.2p_2 = 0$.

Solving the linear system of the above equation and of $p_1 + p_2 = 1$ one finds that $p_1 = 0.4$ [4]. Therefore, the market's share for K in the long run will be 40%.

(ii) The mean number of times that a costumer buys a competitive product between two successive purchases of K is equal to $m_{21} = \frac{p_2}{p_1} = \frac{6}{4} = 1.5$.

4. ***Example:*** In an industry the production of a certain product depends upon the existing stock at the end of each day. Namely, if there exist unsatisfied orders or the stock is zero, then the production of the next day covers the unsatisfied orders plus two more metric units (m.u.). On the contrary, if there exists a non zero stock, there is no production for the next day. We further know that the consumers' demand for the product is either 1 m.u. per day with probability 60%, or 2 m. u. per day with probability 40%. Find the probability to have unsatisfied orders in the long run.

Solution: Since the maximum product's demand is 2 m. u., the production of the factory at the first day of its function must be 2 m. u. and therefore at the end of the day the stock will be either zero or 1 m. u. In the former case the process is repeated in the same way. In the latter case the production of the next day is zero and therefore at the end of the day the stock will be either zero (in this case the process is repeated again in the same way), or there are unsatisfied orders of 1 m. u. In the last case the production of the next day is 3 m. u., i.e. 1 m. u. to cover the unsatisfied orders of the previous day plus 2 m. u., and so on. It becomes therefore evident that, according to the above rhythm of production, there are three possible situations at the end of each day: s_1 = unsatisfied orders of 1 m.u., s_2 = zero stock and s_3 = stock of 1 m.u.

Evidently our problem can be described with an EMC having as states the above possible situations s_i, $i = 1,2,3$. Using the given data it is easy to observe that the transition matrix of the EMC is

$$A = \begin{array}{c} \\ s_1 \\ s_2 \\ s_3 \end{array} \begin{array}{c} \begin{array}{ccc} s_1 & s_2 & s_3 \end{array} \\ \begin{bmatrix} 0 & 0.4 & 0.6 \\ 0 & 0.4 & 0.6 \\ 0.4 & 0.6 & 0 \end{bmatrix} \end{array}.$$

Let $P = [p_1\ p_2\ p_3]$ be the limiting probability vector, then the equality $P = PA$ gives that $p_1 = 0.4p_3$, $p_2 = 0.4p_1 + 0.4p_2 + 0.6p_3$, and $p_3 = 0.6p_1 + 0.6p_2$. Adding the first two of the above equations we find the third one.

Solving the linear system of the first two equations and of $p_1 + p_2 + p_3 = 1$ one finds that $p_1 = 0.15$. Therefore the probability to have unsatisfied orders in the long run is 15%.

ABSORBING MARKOV CHAINS (AMCs)

A state of a MC is called ***absorbing*** if, once entered, it cannot be left. Further a MC is said to be an ***Absorbing MC (AMC)*** if it has at least one absorbing state and if from every state it is possible to reach an absorbing state, not necessarily in one step.

In case of an AMC with k absorbing states, $1 \leq k < n$, we bring its transition matrix A to its ***canonical*** (or ***standard***) ***form*** A^* by listing the absorbing states first and then we make a partition of A^* of the form A^* =

$\begin{bmatrix} I & | & O \\ - & | & - \\ R & | & Q \end{bmatrix}$, where I is the unitary k X k matrix, O is a zero matrix, R is the $(n-k)$ X k transition matrix from the non absorbing to the absorbing states and Q is the $(n-k)$ X $(n-k)$ transition matrix between the non absorbing states.

Denote by I_{n-k} the unitary (n – k) X (n – k) matrix, then it can be shown that the matrix I_{n-k} - Q is always invertible [5]. Then, the ***fundamental matrix*** of the AMC is defined to be the matrix

$$N = [n_{ij}] = (I_{n-k} - Q)^{-1} = \frac{1}{D\,(I_{n-k} - Q)}\, adj\,(I_{n-k} - Q) \qquad (4),$$

where $(I_{n-k} - Q)^{-1}$, $D\,(I_{n-k} - Q)$ and $adj\,(I_{n-k} - Q)$ denote the inverse, the determinant and the ***adjoin*** of the matrix $I_{n-k} - Q$ respectively. We recall that the
$adj\,(I_{n-k} - Q)$ is the matrix of the algebraic complements of the transpose matrix of the matrix $I_{n-k} - Q$ (Morris, 1978, Section 2.4).

It is well known (Kemeny & Snell, 1976, Chapter 3) that *the element n_{ij} of the fundamental matrix N gives the mean number of times in state s_i before the absorption, when the starting state of the AMC is s_j* (where s_i and s_j are non absorbing states).

The above results are illustrated by the following example:

6. Example: An agricultural cooperative applies the following steps for the collection and shelling of a product: s_1 = collection, s_2 = sorting - refining, s_3 = packing and s_4 = shelling. The past experience shows that there is a 20% probability that the quality of the collected product is not satisfactory. In this case the collected quantity is abandoned and a new collection is attempted. It is also known that the duration of each of the stages s_i, i =1, 2, 3, 4, is on average 10, 4, 3 and 45 days respectively. Find the mean time needed for the completion of the whole process.

Solution: The above process can be represented by a finite MC having as states the stages s_i, for the collection and shelling of the product, *i=1, 2, 3, 4*. This MC is obviously an AMC with s_4 being its unique absorbing state. It is straightforward to check that the transition matrix of the AMC is:

$$A=\begin{array}{c|cccc} & s_1 & s_2 & s_3 & s_4 \\ \hline s_1 & 0 & 1 & 0 & 0 \\ s_2 & 0,2 & 0 & 0,8 & 0 \\ s_3 & 0 & 0 & 0 & 1 \\ s_4 & 0 & 0 & 0 & 1 \end{array}$$

. Therefore, the canonical form of A is

$$A^*=\begin{array}{c|c|ccc} & s_5 & s_1 & s_2 & s_3 \\ \hline s_4 & 1 & 0 & 0 & 0 \\ & - & - & - & - \\ s_1 & 0 & 0 & 1 & 0 \\ s_2 & 0 & 0,2 & 0 & 0,8 \\ s_3 & 1 & 0 & 0 & 0 \end{array}.$$

Thus $I_3 - Q = \begin{bmatrix} 1 & -1 & 0 \\ -0.2 & 1 & -0.8 \\ 0 & 0 & 1 \end{bmatrix}$ and applying formula (4) one finds, after a straightforward calculation [6], that

$$N = \begin{array}{c|ccc} & s_1 & s_2 & s_3 \\ \hline s_1 & 1.25 & 1.25 & 1 \\ s_2 & 0.25 & 1.25 & 1 \\ s_3 & 0 & 0 & 1 \end{array}.$$

Therefore, since the AMC always starts from state s_1, the mean number of times in states s_1 and s_2 before the absorption are 1.25 and in state s_3 is 1. Therefore, the mean time needed for the completion of the whole process is 1.25 *(10+4) + 3 + 45 = 65.5 days.

When an AMC has more than one absorbing states, then *the element b_{ij} of the matrix $B = NR = [b_{ij}]$ gives the probability for the AMC starting in state s_i to be absorbed in state s_i* (Kemeny & Snell, 1976, Chapter 3). This is illustrated in the following example, which is a special case of a general problem known as the ***"random – walk"*** problem:

*7. **Example:*** A supermarket has three storehouses, say A_1, A_2 and A_3 between two cities, say C_1 and C_2 , as it is shown in the below diagram:

C_1 ------- A_1 ------- A_2 ------- A_3 ------- C_2

For the delivery of the goods, a truck starts its route every day from one of the storehouses and terminates it in one of the cities. The truck moves each time one place to the right or to the left with the same probability. Find the mean number of stops of the truck to each storehouse during its route and the probability to terminate its route in the city C_1, when it starts it from storehouse A_2.

Solution: We introduce a 5-state MC having the following states:

s_1 (s_5) = the truck arrives to the city C_1 (C_2)

s_2 (s_3, s_4) = the truck arrives to the storehouse A_1 (A_2, A_3).

Obviously the above MC is an AMC and s_1, s_5 are its absorbing states. The canonical form of its transition matrix is:

$$A^* = \begin{array}{c} \\ s_1 \\ s_5 \\ \\ s_2 \\ s_3 \\ s_4 \end{array} \begin{array}{c} \begin{array}{ccccc} s_1 & s_5 & s_2 & s_3 & s_4 \end{array} \\ \left[\begin{array}{cc|ccc} 1 & 0 & 0 & 0 & 0 \\ 0 & 1 & 0 & 0 & 0 \\ - & - & - & - & - \\ 0.5 & 0 & 0 & 0.5 & 0 \\ 0 & 0 & 0,5 & 0 & 0.5 \\ 0 & 0.5 & 0 & 0.5 & 0 \end{array}\right] \end{array}.$$

It is straightforward to check that the fundamental matrix of the AMC is

$$N = (I_3-Q)^{-1} = \begin{array}{c} \\ s_2 \\ s_3 \\ s_4 \end{array} \begin{array}{c} \begin{array}{ccc} s_2 & s_3 & s_4 \end{array} \\ \begin{bmatrix} 1.5 & 1 & 0.5 \\ 1 & 2 & 1 \\ 0.5 & 1 & 1.5 \end{bmatrix} \end{array}$$

Thus, since the truck starts its route from the storehouse A_2 (state s_3), the mean number of its stops to the storehouseA_1 (state s_2) is 1, to the storehouse A_2 (state s_3) is 2 and to the storehouse A_3 (state s_4) is 1.

$$\text{Further, } B = NR = \begin{bmatrix} 1.5 & 1 & 0.5 \\ 1 & 2 & 1 \\ 0.5 & 1 & 1.5 \end{bmatrix} \begin{bmatrix} 0.5 & 0 \\ 0 & 0 \\ 0 & 0.5 \end{bmatrix} = \begin{array}{c} \\ s_2 \\ s_3 \\ s_4 \end{array} \begin{array}{c} \begin{array}{cc} s_1 & s_2 \end{array} \\ \begin{bmatrix} 0.75 & 0.75 \\ 0.5 & 0.5 \\ 0.25 & 0.25 \end{bmatrix} \end{array}$$

Therefore, the probability for the truck to terminate its route to the city C_1 (state s_1), when it starts it from store A_2 (state s_3) is 50%.

Exercise: Solve the previous problem under the hypothesis that the truck moves each time one place to the right with probability p, or one place3 to the left with probability q, for any positive real numbers p and q such that $p + q = 1$, $p \neq q$.

The next example illustrates the fact that a great care is needed sometimes for "translating" correctly the mathematical results of the MC model in terms of the corresponding real situation. This is actually a general principle of the MM process.

8. Example: In a college the minimum period of studies is four years. The statistical analysis has shown that there is a 20% probability for each student to be withdrawn due to unsatisfactory performance and a 30% probability to repeat the same year of studies. Find the probability for a student to graduate, the mean time needed for the graduation and the mean time of his/her attendance in each year of studies.

Solution: We introduce a finite MC with the following states:

s_i = attendance of the i – th year of studies, $i = 1, 2, 3, 4,$

s_5 = withdrawal from the college and

s_6= graduation.

Obviously we have an AMC, where s_5 and s_6 are its absorbing states. The canonical form of its transition matrix is

$$A^* = \begin{array}{c} \\ s_5 \\ s_6 \\ \\ s_1 \\ s_2 \\ s_3 \\ s_4 \end{array} \begin{array}{c} \begin{array}{cccccc} s_5 & s_6 & s_1 & s_2 & s_3 & s_4 \end{array} \\ \left[\begin{array}{cc|cccc} 1 & 0 & 0 & 0 & 0 & 0 \\ 0 & 1 & 0 & 0 & 0 & 0 \\ - & - & - & - & - & - \\ 0.2 & 0 & 0.3 & 0.5 & 0 & 0 \\ 0.2 & 0 & 0 & 0.3 & 0.5 & 0 \\ 0.2 & 0 & 0 & 0 & 0.3 & 0.5 \\ 0.2 & 0 & 0 & 0 & 0 & 0.3 \end{array}\right] \end{array}$$

Utilizing a common software package [7] it is straightforward to check that the fundamental matrix of the chain is

$$N = (I_4 - Q)^{-1} = \begin{array}{c} \\ s_1 \\ s_2 \\ s_3 \\ s_4 \end{array} \begin{array}{c} \begin{array}{cccc} s_1 & s_2 & s_3 & s_4 \end{array} \\ \begin{bmatrix} 1.429 & 1.02 & 0.729 & 0.521 \\ 0 & 1.429 & 1.02 & 0.729 \\ 0 & 0 & 1.429 & 1.02 \\ 0 & 0 & 0 & 1.429 \end{bmatrix} \end{array}$$

.

$$\text{Therefore, } B=NR = \begin{array}{c} \\ s_1 \\ s_2 \\ s_3 \\ s_4 \end{array} \begin{array}{c} \begin{array}{cc} s_5 & s_6 \end{array} \\ \begin{bmatrix} 0.74 & 0.261 \\ 0.636 & 0.365 \\ 0.49 & 0.51 \\ 0.286 & 0.715 \end{bmatrix} \end{array}..$$

Observing the fundamental matrix *N* of the chain, one finds that n_{13} = 0.729 and n_{14} = 0.521, which means that for a first year student the mean time of attendance in the third and fourth year of studies is less than one year! However this is not embarrassing, because there is always a possibility for a student to be withdrawn due to unsatisfactory performance before entering the third, or fourth, year of studies.

Since $n_{11} = n_{22} = n_{33} = n_{44} = 1.429$, one finds that the mean time of attendance of a student in each year of studies is 1.429 years, while the mean time needed for his/her graduation is 1.429 * 4 = 5.716 years. Further, observing the matrix *B* one finds that $b_{15} = 0.74$, i.e. the probability of a student to graduate is 74%.

We close this section with the following unsolved problem that enables the reader to check his/her ability for solving mathematical modelling problems involving the use of MCs:

9. Problem: In an island the daily weather is either rainy (R), or cloudy (C), or bright (B). It is statistically known that, if the weather is rainy the probability to be also rainy during the next day is 20%, whereas the probability to be bright is 50%. It is also known that if the weather is cloudy the probability to be also cloudy during the next day is 50% and the probability to be bright is 20%. Finally, if the weather is bright the probability to be also bright during the next day is 10% while the probability to be rainy is 70%.

(i) Find the probability for the weather to be cloudy next Sunday, and

(ii) If it is cloudy on Sunday, to be rainy on Tuesday.

CONCLUSIONS

The following conclusions can be drawn from the discussion performed in this chapter:

- The theory of MCs is a successful combination of Linear Algebra and Probability theory, which enables one to make short and long run forecasts for the evolution of various phenomena of the real world, which can be approximately characterized as having a ***one step memory*** (Markov property).

- The ***short run*** forecasts are obtained, regardless the type of the MC, by calculating its ***transition matrix*** and the ***probability vector*** of the corresponding step (general form of the MC model). On the contrary, the ***long run*** forecasts (equilibrium situation of the MC) are obtained, in case of ***EMCs*** only, by calculating the corresponding ***limiting probability vector***.

- In case of an ***AMC*** one proceeds to the study of the corresponding problem by writing the transition matrix in its ***canonical form*** and calculating the ***fundamental matrix*** N of the AMC. The entries of N give the mean number of times in each non absorbing state before absorption, for each possible starting non absorbing state. When an AMC has more than one absorbing states, then the matrix $B = NR$, where R is the ***transition matrix from the non absorbing to the absorbing states***, enables one to calculate the probabilities for the chain to reach a certain absorbing state, when it starts from a certain non absorbing state.

REFERENCES

Bartholomew, D.J. (1973), *Stochastic Models for Social Processes*, J. Wiley and Sons, London.

Kemeny, J. G. & Snell, J. L. (1963), *Mathematical Models in the Social Sciences*, Ginn and Company, New York, USA.

Kemeny, J. G. & Snell J. L. (1976), *Finite Markov Chains*, Springer - Verlag, New York, USA.

Suppes, P. & Atkinson, R. C. (1960), *Markov Learning Models for Multiperson Interactions*, Stanford University Press, Stanford-California, USA.

Mathcad 2001i (2001), *User's Guide with Reference Manual,* Mathsoft, Cambridge, MA, USA

Morris, A. O. (1978), *An Introduction to Linear Algebra,* Van Nostrand Beinhold Company Ltd., Berkshire, England.

Rozanov, Y. A. (1972), *Probability Theory: A Concise Course* (Revised English Edition Translated and Edited by R. A. Silverman), Dover Publications, New York, USA.

Suppes, P. & Atkinson, R. C. (1960), *Markov Learning Models for Multiperson Interactions*, Stanford University Press, Stanford-California, USA.

Voskoglou, M.Gr. (2016), Applications of Finite Markov Chain Models to Management, *American Journalof Computational and Applied Mathematics*, 6(1), 7-13.

ENDNOTES

[1] According to the strict mathematical definition the Markov property requires a one step memory ONLY. However, this is not so easy to happen in practice, which makes many authors to give the weaker (approximate) definition presented here. In this way the theory of MCs can be applied for the modelling and solution of much more real world problems.

[2] Let $B_1, B_2, \ldots, B_n$ be a finite number of mutually exclusive events (i.e. $B_i \cap B_j = \varnothing$, if $i \neq j$). Assume further that the event A always occurs together with only one of the events $B_1, B_2, \ldots, B_n$.

Then $A = (B_1 \cap A) \cup (B_2 \cap A) \cup \ldots \cup (B_n \cap A)$. But $(B_i \cap A) \cap (B_j \cap A) = \varnothing$, if $i \neq j$, therefore $P(A) = P(B_1 \cap A) + P(B_2 \cap A) + \ldots + P(B_n \cap A)$. Further, the conditional probability $P(A|B_i) = \frac{P(A \cap B_i)}{P(B_i)}$.

Hence, the previous equality gives that $P(A) = \sum_{i=1}^{n} P(B_i) P(A \mid B_i)$.

The last equation is known as the ***total probability formula.***

[3] It is recalled that if $A = [a_{ij}]$ is a n x m matrix and $B = [b_{jk}]$ is a m x s matrix, then their ***product*** AB is defined to be the n x s matrix $C = [c_{ik}]$, such that

$c_{ik} = \sum_{j=1}^{m} a_{ij} b_{jk}$. In our case A = $\begin{bmatrix} 0.7 & 0.3 \\ 0.2 & 0.8 \end{bmatrix}$, therefore $A^2 = AA = \begin{bmatrix} 0.55 & 0.45 \\ 0.3 & 0.7 \end{bmatrix}$ and

$P_2 = [0\ 1]\ A^2 = [0.3\ \ 0.7]$.

[4] It is recalled that a standard way to solve the linear system is to apply

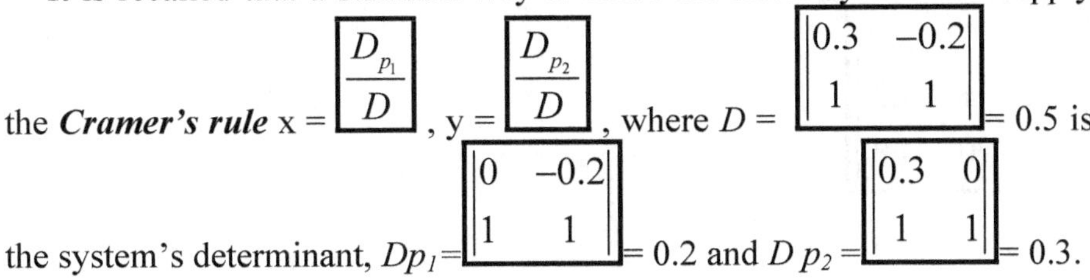

the ***Cramer's rule*** $x = \frac{D_{p_1}}{D}$, $y = \frac{D_{p_2}}{D}$, where $D = \begin{vmatrix} 0.3 & -0.2 \\ 1 & 1 \end{vmatrix} = 0.5$ is the system's determinant, $Dp_1 = \begin{vmatrix} 0 & -0.2 \\ 1 & 1 \end{vmatrix} = 0.2$ and $D\,p_2 = \begin{vmatrix} 0.3 & 0 \\ 1 & 1 \end{vmatrix} = 0.3$.

[5] It is easy to check that $(A^*)^m = \begin{bmatrix} I & | & O \\ - & | & - \\ R^* & | & Q^m \end{bmatrix}$, for all positive integers *m*, where R* is a matrix that it is not calculated here. But the MC will be absorbed after a finite number of steps and *Q* is the transition matrix between its non absorbing states, which means that there exists a positive integer *s* such that $Q^s = 0$. Therefore $I_{n-k} = I_{n-k} - Q^s = (I_{n-k} - Q)(I_{n-k} + Q + Q^2 + \ldots.. + Q^{s-1})$, which shows that

$(I_{n-k} - Q)^{-1} = I_{n-k} + Q + Q^2 + \ldots.. + Q^{s-1}$.

[6] In fact, $D(I_3 - Q) = \begin{vmatrix} 1 & -0.8 \\ 0 & 1 \end{vmatrix} + \begin{vmatrix} -0.2 & -0.8 \\ 0 & 1 \end{vmatrix} = 0.8$, $(I_3 - Q)^t = \begin{bmatrix} 1 & -0.2 & 0 \\ -1 & 1 & 0 \\ 0 & -0.8 & 1 \end{bmatrix}$,

(the transpose matrix of $I_3 - Q$) and

$adj\ (I_3 - Q)$ = 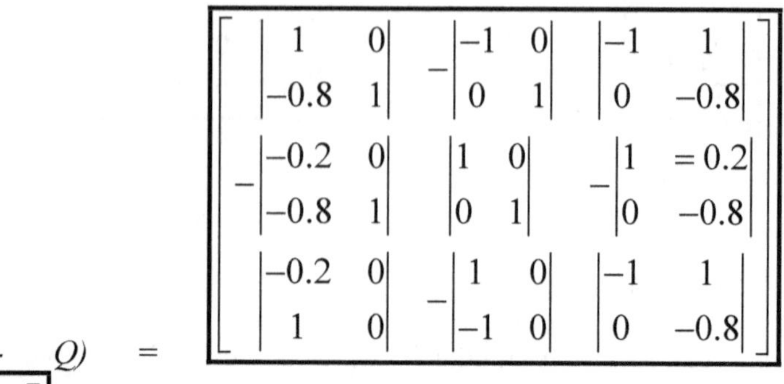 = $\begin{bmatrix} 1 & 1 & 0.8 \\ 0.2 & 1 & 0.8 \\ 0 & 0 & 0.8 \end{bmatrix}$.

Therefore, formula (4) gives that $N = \frac{1}{0.8} \begin{bmatrix} 1 & 1 & 0.8 \\ 0.2 & 1 & 0.8 \\ 0 & 0 & 0.8 \end{bmatrix} = \begin{bmatrix} 1.25 & 1.25 & 1 \\ 0.25 & 1.25 & 1 \\ 0 & 0 & 1 \end{bmatrix}$.

[7] For example, utilizing the ***Mathcad*** (2001) software to find the inverse of a given matrix A one clicks on the desktop, where it appears a red cross. Then he/she writes "A=" and he/she clicks on View → Toolbars → Matrix. In the window "Matrix" he/she clicks on the existing figure of the matrix and he/she writes the numbers of rows and columns of A in the icon appearing on the desktop. Then he/she writes the entries of A and in the window "Matrix" he/she clicks on X^{-1} . He/she writes "A" in the blank space on the desktop and then he/she presses the key "="to get the required result.

To calculate a power of A one must open the window "Calculator" through View → Toolbars → Calculator. Then he/she clicks on X^n, he/she writes the required value of n and he/she presses the key "="to get the required result. In an analogous way one can calculate the product of two matrices and make any other operations between matrices.

CHAPTER 3

Finite Markov Chain Models in Learning Contexts

ABSTRACT

Finite MC models are developed in this Chapter for Decision-Making, for Problem-Solving, for the description of Mathematical Modelling in the classroom, for assessing the Analogical Reasoning skills and for evaluating the effectiveness of Case – Based Reasoning systems. The chapter closes with a brief reference to other MC applications developed in earlier works and with a general conclusion about the advantages and disadvantages of the MC models in representing human activities.

AN AMC FOR DECISION-MAKING

The DM process

Decision-Making (DM) is the process of choosing a solution between two or more alternatives, aiming to achieve the best possible results for a given problem. Obviously the above process has sense if, and only if, there exist more than one feasible solutions, together with a suitable criterion (or criteria) that helps the decision maker (d-m) to choose the best among these solutions. The choice of the suitable criterion, especially when the results of DM are affected by random events, depends upon the desired goals of the d-m; e.g. optimistic or conservative criterion etc.

The rapid technological progress during the last 60-70 years, the enormous changes happened to the local and international societies and other similar reasons made in many cases the DM process a very difficult task, which is impossible to be based on the d-m's experience, intuition and skills only, as it usually happened in the past. Thus, from the beginning of the 1950's a progressive development started of a systematic methodology for the DM process, which is based on Probability Theory, Statistics, Economics, Psychology, etc. and it is known as ***Statistical Decision Theory*** (Berger, 1980)).

According to the nowadays existing standards the DM process involves the following steps:

- d_1: Analysis of the decision-problem, i.e. understanding, simplifying and reformulating the problem in a way that enables the application of the DM process on it.

- d_2: Collection and interpretation of all the necessary information related to the problem.

- d_3: Determination of all the alternative feasible solutions.

- d_4: Choice of the best solution in terms of the suitable, according to the d-m's goals and targets, criterion.

One could add one more step to the above process, the decision's ***verification*** according to the results obtained by applying the decision in practice. However, this step is extended to areas, which, due to their depth and importance for management, have become autonomous. Therefore, the step of verification is usually examined separately from the other steps of the DM process.

The first three steps of DM are continuous in the sense that the completion of each one of them usually needs some time, during which the d-m's reasoning is characterized by transitions between hierarchically neighbouring steps. The ***flow-diagram*** of the DM process is represented in Figure 1 below:

$$d_1 \boxed{\leftrightarrow} d_2 \boxed{\leftrightarrow} d_3 \boxed{\rightarrow} d_4$$

Figure 1: The flow-diagram of the DM process

The MC model for DM

In Voskoglou (2014) we introduced a 4-state MC on the steps d_i, i = 1, 2, 3, 4, of the DM process, in which d_1 is always the starting state. Obviously this is an AMC, with d_4 being its unique absorbing state.

Let us denote by p_{ij} the transition probability from state d_i to d_j , i, j = 1, 2, 3 ,4. Then, the transition matrix of the MC is

$$A = \begin{array}{c|cccc} & d_1 & d_2 & d_3 & d_4 \\ \hline d_1 & 0 & 1 & 0 & 0 \\ d_2 & p_{21} & 0 & p_{23} & 0 \\ d_3 & 0 & p_{32} & 0 & p_{34} \\ d_4 & 0 & 0 & 0 & 1 \end{array},$$

with $p_{21}+p_{23} = p_{32}+p_{34} = 1$.

Let us further denote by $\varphi_0, \varphi_1, \varphi_2$, ..., the successive phases (steps) of the MC and let $P_i = [p_1^{(i)}\ p_2^{(i)}\ p_3^{(i}\ p_4^{(i)}]$ be its probability vector at phase φ_i , i = 0, 1, 2,.... . Then, since d_1 is always the starting state, we have

$P_0 = [1\ 0\ 0\ 0]$, $P_1 = P_0A = [0\ 1\ 0\ 0]$, $P_2 = P_1 A = [p_{21}\ 0\ p_{23}\ 0]$, $P_3 = P_2 A =$ $= [0\ \ p_{21}+ p_{23}p_{32}\ \ 0\ \ p_{23}p_{34}]$, $P_4 = P_3A = [p_{21}^2+p_{21}p_{23}p_{32}\ \ 0\ \ p_{21}p_{23}+ p_{23}^2 p_{32}\ \ p_{23}p_{34}]$ and so on.

The canonical form of A is

$$A^* = \begin{array}{c|c:ccc} & d_4 & d_1 & d_2 & d_3 \\ \hline d_4 & 1 & 0 & 0 & 0 \\ \hdashline d_1 & 0 & 0 & 1 & 0 \\ d_2 & 0 & p_{21} & 0 & p_{23} \\ d_3 & p_{34} & 0 & p_{32} & 0 \end{array}.$$

Therefore, the transition matrix among the non absorbing states is

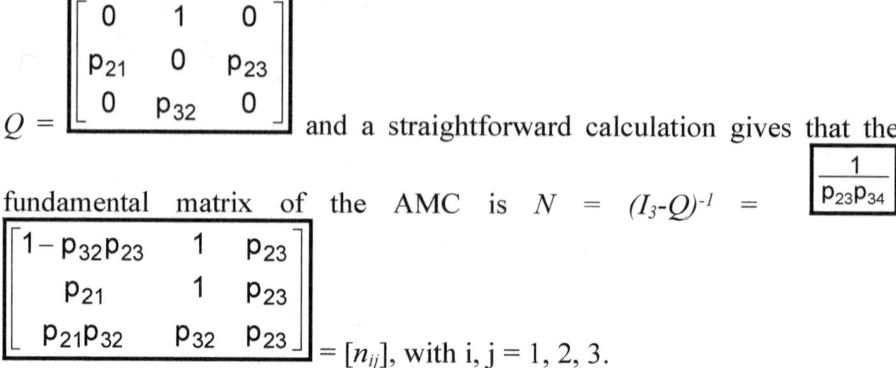

$Q = \begin{bmatrix} 0 & 1 & 0 \\ p_{21} & 0 & p_{23} \\ 0 & p_{32} & 0 \end{bmatrix}$ and a straightforward calculation gives that the fundamental matrix of the AMC is $N = (I_3-Q)^{-1} = \frac{1}{p_{23}p_{34}} \begin{bmatrix} 1-p_{32}p_{23} & 1 & p_{23} \\ p_{21} & 1 & p_{23} \\ p_{21}p_{32} & p_{32} & p_{23} \end{bmatrix} = [n_{ij}]$, with i, j = 1, 2, 3.

We recall that the ij-th entry of N gives the mean number of times in state d_j before the absorption, when the chain is started in state d_i. Therefore, since in our case d_1 is always the starting state, the mean number of steps taken before absorption is given by $t = \sum_{i=1}^{3} n_{1i} = \frac{2+p_{23}p_{34}}{p_{23}p_{34}}$ (1).

Obviously, the bigger the value of t, the more the difficulties faced during the DM process; in other words t provides an indication for the difficulty of the DM process (another indication could be the time spent by the d-m to make the decision, etc.).

Example

A company A, which produces and trades a certain product, say K, is searching for the proper place to build a new factory. The company's management wants to calculate the mean number of steps required for making the decision about the factory, as well as the probability for the DM process to be terminated in four steps.

The analysis (***step*** d_1 of Section 3.1.1) of the DM problem about the place of the factory has shown that the decision's profitability depends upon the quality of the competitive to the K products produced by the existing in the area similar companies.

The market's research (***step*** d_2) has shown that there is only one similar company in the area, say B, which produces three different types of competitive to K products, say W_1, W_2 and W_3.

The general situation in the area (communications, traffic, the already existing factories and storehouses of the two companies, etc.), combined with the funds available by the company A for the construction of its new factory, suggest four favourable places, say P_1, P_2, P_3 and P_4 for the construction of the new factory (***step*** d_3).

(*d_3* → *d_2*): The market's research has shown that the expected profits of the company A with respect to the favourable places for the construction of its new factory and to the types of product of company B are those presented in Table 1 below:

Table1: Net profits of the company A

	P_1	P_2	P_3	P_4
W_1	3	8	5	4
W_2	4	2	6	5
W_3	2	1	1	–1

(*d_2* → *d_3*): From Table 1 it becomes evident that the feasible solution P_4 is worse than P_3 and therefore P_4 is rejected.

The management of the company does not want to risk of earning low profits by the new factory, which means that it must adopt a conservative criterion for the choice of the best place for building it. A frequently used in such cases criterion is the Wald's criterion (Berger, 1980), which is based on Murphy's law assuming that the worst possible fact to be happen will finally happen. This criterion suggests to maximize the minimal possible for each case profits ***(maximin of payoffs)***. In other words, since the minimal expected profit from the choice of P_1 is 2 monetary units and from the choice of P_2 or of P_3 is 1 monetary unit (see Table 1), the place P_1 must be chosen for building the new factory (***step*** d_4).

Application of the MC model: From the above statement it becomes evident that $p_{21} = 0$ and $p_{23} = 1$. We shall also show that $p_{32} = p_{34} = 0.5$. In fact, when the MC reaches the state d_3 for first time, the probability of returning to d_2 at the next step is 1, since the collection and interpretation of new information is necessary. On the contrary, the second time that the MC reaches d_3 the probability of returning to d_2 at the next step is 0, since no more information is needed for the choice of the best solution. Therefore the transition probability p_{32} is equal to the mean value $\frac{0+1}{2}$ = 0.5 and also $p_{34} = 1 - p_{32} = 0.5$.

Replacing the above values to equation (1) of Section 3.1.2 one finds that

$P_3 = [0\ \ 0.5\ \ 0\ \ 0.5]$. Therefore, $p_4^{(3)} = 0.5$, which means that the probability for the DM process to be terminated in 4 steps is 50%. This could happen, if there was no feasible solution worse than one of the others and therefore we didn't reject any of them, as we did above for P_4 (Table 1).

Further, it is straightforward to check that $N = \frac{1}{0.5}\begin{bmatrix} 0.5 & 1 & 1 \\ 0 & 1 & 1 \\ 0 & 0.5 & 1 \end{bmatrix}$, wherefrom one finds that $n_{11} = 1$, $n_{12} = n_{13} = 2$. Thus, the mean number of steps for the DM process before taking the decision is t = 5 steps.

MEASURING PROBLEM-SOLVING SKILLS

Problem-Solving in Mathematics Education

As the human society moved from an industrial to a knowledge society, it can be argued that the nature of many problems has been changed and new problems have arisen which may require a different approach to overcome them. Educational institutions and governments have recognized long ago the importance of ***Problem–Solving (PS)*** and volumes of research have been written about it. Universities and other higher learning institutions are entrusted with the task of producing graduates that have higher order PS skills among other skills.

Mathematics by its nature is a subject whereby PS forms its essence. In Voskoglou (2011, 2016) we have examined the role of the problem in learning mathematics and we have attempted a review of the evolution of research on PS in mathematics education from the time that Polya presented his first ideas on the subject until today. Here is a rough chronology of that progress:

1950's – 1960's: Introduction of the ***heuristic strategies*** by Polya (1945, 1954, 1962/65, 1963, 1973, etc.) for teaching the PS process.

1970's: Emergency of mathematics education as a self – sufficient science. The research on PS was still based on Polya's ideas, while the research methods were almost exclusively statistical.

1980's: A framework appeared describing the PS process, and reasons for success or failure in PS, e.g. Schoenfeld's (1980) ***Expert Performance Model (EPM)***, etc.

1990's: Models of teaching mathematics were developed using PS, e.g. constructivist view of learning (von Claserfeld, 1987, etc.), MM and applications (e.g. Voskoglou 2015), etc.

2000's - today: In contrast to the earlier work on PS, which was focused mainly on analyzing the PS process and on describing the proper heuristic strategies to be used in each of its stages, the research has turned mainly on solvers' behavior and required attributes during the PS process; e. g. ***Multidimensional PS Framework (MPSF)*** of Carlson and Bloom (2005), Schoenfeld's theory of ***goal-directed behavior*** (2010), etc.

Carlson and Bloom (2005) drawing from the large amount of literature related to PS developed a broad taxonomy to characterize major PS attributes that have been identified as relevant to PS success. This taxonomy gave genesis to their MPSF model, which includes the following steps: ***Orientation, Planning, Executing*** and ***Checking***. It has been observed that once the solvers oriented themselves to the problem space, the plan-execute-check cycle was usually repeated through out the remainder of the solution process; only in a few cases a solver obtained the solution of a problem by making this cycle only once. Thus embedded in the framework are two cycles, one cycling forward and one cycling backward, each of which includes the three out of the four steps, i.e. planning, executing and checking. It has been also observed that, when contemplating various solution approaches during the planning step of the PS process, the solvers were at times engaged in a ***conjecture-imagine-evaluate (accept/reject)*** sub-cycle. Therefore, apart of the two main cycles, embedded in the MPSF is also the above sub-cycle, which is connected to the step of planning (Carlson & Bloom, 2005, Figure 1).

Note that the main phases of the MPSF are actually the same to the steps of Schoenfeld's (1980) EPM for PS; only their names are stated differently. In fact, a careful inspection of the two PS models shows that Orientation (***S_1***) corresponds to Schoenfeld's ***Analysis*** of the problem, Planning (***S_2***) corresponds to the ***Design*** of the solution, the conjecture-imagine-evaluate sub-cycle (***S_3***) corresponds to Schoenfeld's step of ***Exploration***, Executing (***S_4***) corresponds to the ***Implementation*** of the solution and finally Checking (***S_5***) corresponds to Schoenfeld's ***Verification*** of the solution (Voskoglou, 2011b). However, there exists a basic qualitative difference between the two models: While in the MPSF the emphasis is turned to the solver's behavior and required attributes, the EPM is oriented towards the PS process itself describing the proper heuristic strategies that may be used at each step of the PS process.

The flow-diagram of the PS process is shown in Figure 2. In fact, a solver who faces difficulties at step ***S_2***, proceeds to ***S_3***. From there, if the difficulties are surpassed, he/she returns to ***S_2*** to continue the PS process. Otherwise he/she returns to the staring step ***S_1*** searching for additional information from problem's data that possibly has been elapsed at first glance. The same circle may be repeated several times.

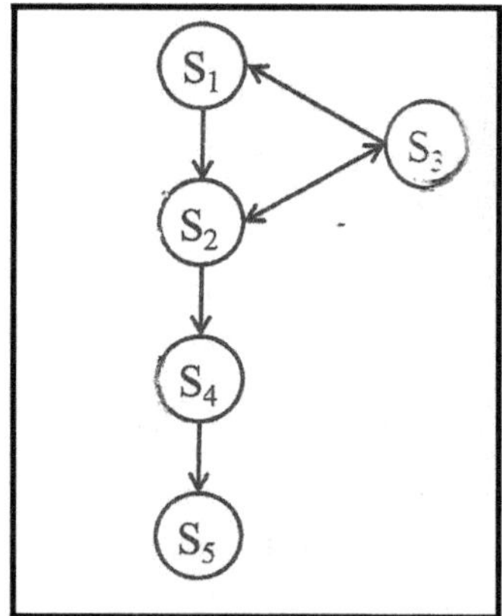

Figure 2: The flow-diagram of the PS process

An AMC Model for PS

In earlier works (Voskoglou & Perdikaris, 1991, 1993) we have developed with the help of the diagram of Figure 5, an AMC model for the PS process as follows:

We introduce a finite MC on the steps $\boldsymbol{S_i}$, i=1, 2,…,5 of the PS process. Obviously this MC is an AMC with $\boldsymbol{S_5}$ being its unique absorbing state. Further, it is easy to check that its transition matrix has the form

$$A=\begin{array}{c|ccccc|} & S_1 & S_2 & S_3 & S_4 & S_5 \\ S_1 & 0 & 1 & 0 & 0 & 0 \\ S_2 & 0 & 0 & p_{23} & p_{24} & 0 \\ S_3 & p_{31} & p_{32} & 0 & 0 & 0 \\ S_4 & 0 & 0 & 0 & 0 & 1 \\ S_5 & 0 & 0 & 0 & 0 & 1 \end{array}, \text{ with } p_{23}+p_{24}=p_{31}+p_{32}=1.$$

Let us denote by $\varphi_0, \varphi_1, \varphi_2, \ldots\ldots$, the successive phases of the MC and let $P_i = [p_1^{(i)}\ p_2^{(i)}\ p_3^{(i}\ p_4^{(i)}\ p_5^{(i)}]$ be its probability vector at phase φ_i , $i = 0, 1, 2, \ldots.$. Then, since $\boldsymbol{S_1}$ is always the starting state, we have $P_0 = [1\ 0\ 0\ 0\ 0]$, $P_1 = P_0 A$
$=$ [0 1 0 0 0], $P_2 = P_1 A$ = [0 0 p_{23} p_{24} 0] and in general P_{n+1}` $= P_n A^n$, for all non negative integers *n.* The last equation enables one to make short run forecasts for the evolution of the PS process by calculating the probabilities of the MC to be in each of its five states at each one of its phases.

It is also easy to observe that the transition matrix among the non absorbing states of the MC is equal to

$$Q=\begin{bmatrix} 0 & 1 & 0 & 0 \\ 0 & p_{22} & p_{23} & p_{24} \\ p_{31} & p_{32} & 0 & 0 \\ 0 & 0 & 0 & 0 \end{bmatrix}.$$

Then the fundamental matrix $N = [n_{ij}]$, $i, j = 1, 2, 3, 4$, of the chain is calculated by $N = (I_4 - Q)^{-1}$, where I_4 denotes the unitary 4X4 matrix. Since the element n_{ij} of N gives the mean number of times at state $\boldsymbol{S_j}$ before the absorption, when the MC starts from $\boldsymbol{S_t}$ and since in our case the MC starts always from $\boldsymbol{S_1}$, it becomes evident that the sum t =

$\sum_{j=1}^{4} n_{1j}$ gives the mean number of phases of the MC before the absorption. Making the necessary calculations one finds that

$$t = \frac{3 - p_{23}p_{32}}{p_{24}} \quad (2).$$

Obviously, the more are the difficulties during the PS process, the bigger is the value of t. Therefore, a smaller value of t is connected to a better solver performance. In other words, the value of t gives an indication of the ability either of different solver groups for solving the same problem, or of the same group for solving different problems. Of course this is not the unique indication; e.g. another one could be the total time spent by the group for the completion of the PS process, etc.

Applications of the AMC Model

The PS process of the following two problems by a group of 40 students of the Graduate TEI of Western Greece illustrates the applicability of the AMC model developed in Section 3.2.2 to real PS situations. The time allowed by the instructor for the solution of each problem was 20 minutes.

Problem 1: Given the matrix $M = \begin{bmatrix} 1 & 2 & 2 \\ 0 & 1 & 2 \\ 0 & 0 & 1 \end{bmatrix}$ and a positive integer n, calculate the matrix M^n.

PS process: In the first 15 minutes 30 students solved the problem. To the rest of them the instructor gave the following hint: "Applying induction on n try to show that $M^n = \begin{bmatrix} 1 & 2n & 2n^2 \\ 0 & 1 & 2n \\ 0 & 0 & 1 \end{bmatrix}$". Then six more students solved the problem in the time allowed for solution.

The transition probabilities involved in the AMC model can be calculated in this case as follows (Figure 3): Initially all solvers proceed from S_1 to S_2. Then 30 of them proceed straightforward through S_4 to the absorbing state S_5. The other 10 solvers proceed from S_2 to S_3. From there, 6 of them return to S_2 and reach S_5 through S_4. The other four solvers return to S_1 and they remain there being unable to make any other movements for the problem's solution.

Therefore, since we have 46 in total "arrivals" to S_2 , 36 in total "departures" from S_2 to S_4 and 10 "departures" from S_2 to S_3 , one finds that $p_{24} = \frac{36}{46}$ and $p_{23} = \frac{10}{46}$. Similarly it turns out that $p_{32} = \frac{6}{10}$ and $p_{31} = \frac{4}{10}$.

Replacing the above values to equation (2) one finds that $t = \frac{132}{36} \approx$ 3.67 steps. Also, as an example of a short run forecast, since $P_2 = [0\ 0\ p_{23}\ p_{24}\ 0]$, the probability for the PS process to be at its third phase in the step of executing is equal to $\frac{36}{46}$, or approximately equal to 78.26%.

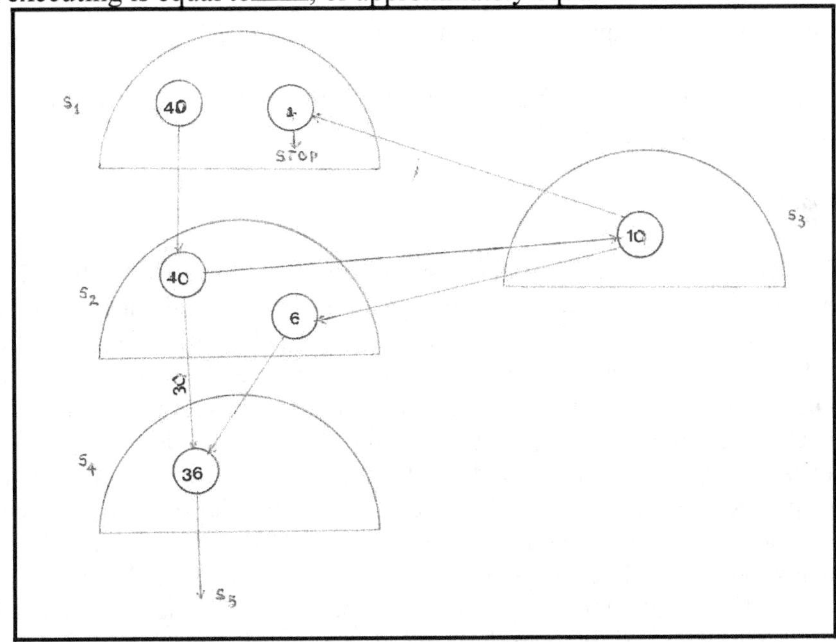

Figure 3: Flow- diagram of solver movements in Problem 1

Problem 2: Let *a*, *b*, *c*, *d* be given numbers between 0 and 1. Prove that

$(1-a)(1-b)(1-c)(1-d) > 1-a-b-c-d$ (Schoenfeld, 1980, problem 2)

PS process: In the first 15 minutes eight students solved the problem as follows: "It is enough to show that $(1-b-a+ab)(1-d-c+cd) > 1-a-b-c-d$, or $ab+ac+ad+bc+(bd+cd+abcd) > abc+abd+acd+bcd$. But $ab > abc$, $ac > acd$, $ad > abd$, $bc > bcd$ and the result follows, since $bd+cd+abcd > 0$ ".

To the rest of the students the instructor gave the hint: "Try first to solve the corresponding problem in two and then in three variables". Then 23 more students solved the problem in the time allowed for solution.

In this case, an argument analogous to that developed in Problem 1 shows that

$p_{24} = \frac{31}{63}$, $p_{23} = \frac{32}{63}$, $p_{32} = \frac{23}{32}$ and $p_{31} = \frac{9}{32}$ and the replacement of the above values to equation (1) gives that $t = \frac{166}{31} \approx 5.35$ steps.

Therefore, although Problem 2 involved elementary Algebra only, the students faced more difficulties than in Problem 1 to solve it.

Remark

The analysis of the PS process of the two problems of Section 3.2.3 shows that the calculation of the transition probabilities involved was based on the description of the solver *assumed behavior,* i.e. on how they could act and not on how they really act in practice for solving the problems. In Problem 1 for example, we have tacitly assumed that the 30 students who solved it in the first 15 minutes, they proceeded linearly from S_2 to S_5 through S_4, which could not be true. In fact, some of them could have passed from S_3 first and possibly they could have made the same circle more than once. Similarly, we have tacitly assumed that the six students who failed to solve the problem proceeded for S_3 to S_1 and remained there until the end of the PS process, which also could not be true. In fact, some of them in their effort to solve the problem could have repeated the same circle more than once. The only way for the instructor to be helped to know if the assumed student behavior can be considered as a reasonable approach to their real behavior is to perform *interviews* asking the students about how they tried to solve the problems. This could be done during a research project, but it is technically difficult - due to the lack of time - to be attempted in the everyday practice, if the instructor decides to use the AMC model for evaluating the student overall performance. However, even in the latter case the student assessment through the AMC model is better providing to the instructor more information than the traditional way of marking the student papers and calculating the mean value of their marks, which is based on student final outcomes only.

Note also that in more complicated problems the calculation of the transition probabilities could be much more difficult, or even impossible.

AN EMC MODEL FOR THE MM PROCESS

The EMC Model

MM appears today as a dynamic tool for teaching and learning mathematics, because it connects mathematics with our everyday life giving the possibility to students to understand its usefulness in practice and therefore increasing their interest about mathematics.

Voskoglou (1995) developed a MC model to describe the MM process in the classroom by introducing an EMC on its main steps. A more detailed version of this model was presented in 2005 at the 12th ICTMA Conference in London (Voskoglou, 2007). The flow-diagram of Voskoglou's model is represented in Figure 4, where the arrows show the possible transitions between the MC's states, which are:

S_1: ***Analysis*** of the problem, i.e. comprehension of its statement and recognition of the restrictions and requirements imposed by the real system.

S_2: ***Mathematization***, i.e. formulation of the problem in a way that it will be ready for mathematical treatment and construction of the model.

S_3: ***Solution*** of the model.

S_4: ***Validation*** (control) of the model, which is usually achieved by reproducing, through the model, the behaviour of the real system under the conditions existing before the solution of the model and by comparing it to the data existing from the past "history" of the corresponding system. In cases of systems having no past history, an extra simulation model could be used for performing the validation of the mathematical model.

S_5: ***Interpretation*** of the final mathematical results and implementation of them to the real system.

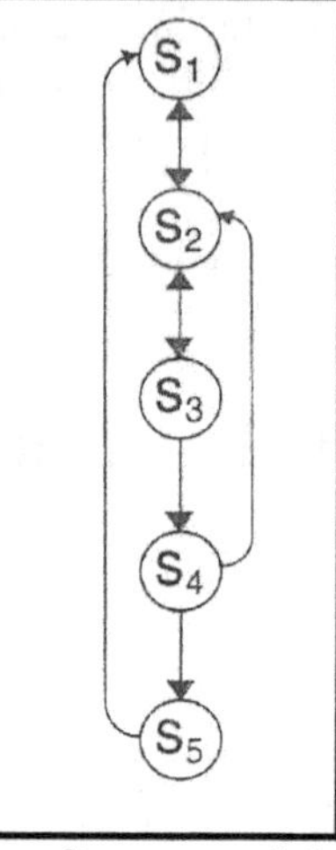

Figure 4: Flow-diagram of the MM process

If the model's control shows that it does not represent correctly the real situation (state **S_4**), then a transfer becomes necessary from **S_4** to **S_2** in order to make the necessary corrections to the model. Finally, when state S_5 is successfully completed, it is assumed that a new MM problem is given for solution by the instructor to the class and therefore the process restarts again from **S_1**. Under the light of the last hypothesis it becomes evident that the constructed MC is an EMC.

The transition matrix of the EMC is

$$A = \begin{array}{c} \begin{array}{ccccc} S_1 & S_2 & S_3 & S_4 & S_5 \end{array} \\ \left[\begin{array}{ccccc} 0 & 1 & 0 & 0 & 0 \\ p_{21} & 0 & p_{23} & 0 & 0 \\ 0 & p_{32} & 0 & p_{34} & 0 \\ 0 & p_{42} & 0 & 0 & p_{45} \\ 1 & 0 & 0 & 0 & 0 \end{array}\right] \end{array} \begin{array}{c} \\ s_1 \\ s_2 \\ s_3 \\ s_4 \\ s_5 \end{array},$$

with $p_{21}+p_{23}=p_{32}+p_{34}=p_{42}+p_{45}=1$ (1).

Let $Q = [\ a_1 \ \ a_2 \ a_3 \ a_4 \ a_5\ ]$ be the EMC's limiting probability vector. Then the equation Q = QA gives that $\alpha_1 = p_{21}\alpha_2 + \alpha_5$, $\alpha_2 = \alpha_1 + p_{32}\,\alpha_3 + p_{42}\,\alpha_4$, $\alpha_3 = p_{23}\alpha_2$, $\alpha_4 = p_{34}\alpha_3$ and $\alpha_5 = p_{45}\alpha_4$. Adding the first four of the above equations and using (1) one finds the fifth equation, which therefore is omitted. Replacing the fifth equation with $\alpha_1+\alpha_2+\alpha_3+\alpha_4+\alpha_5 = 1$ it results to a linear system of five equations with unknowns the α_i's, $i = 1, 2, 3, 4, 5$. Since the system's determinant D = $2p_{34}p_{23}p_{45} + p_{23}p_{34} + 2 > 0$, applying the Cramer's rule one finds the unique solution:

$$\alpha_1 = \frac{1-p_{23}+p_{23}p_{34}p_{45}}{D},\ \alpha_2 = \frac{1}{D},\ \alpha_3 = \frac{p_{23}}{D},\ \alpha_4 = \frac{p_{34}p_{23}}{D},\ \alpha_5 = \frac{p_{34}p_{23}p_{45}}{D} \quad (2).$$

We recall that the a_i's, $i = 1, 2, \ldots, 5$ calculate the probabilities for the EMC to be in each of its states in the long run, which means that they characterize the "*gravity*" of each step of the MM process. Further, since the MM process after the successful completion of step $\mathbf{S_5}$ restarts again from $\mathbf{S_1}$, the sum

$$m = \sum_{i=1}^{4} m_{i5} = \sum_{i=1}^{4} \frac{\alpha_i}{\alpha_5} = \frac{\sum_{i=1}^{4}\alpha_i}{\alpha_5} = \frac{1-\alpha_5}{\alpha_5} \quad (2)$$

calculates the mean number of steps between two successive occurrences of $\mathbf{S_5}$. It becomes therefore evident that the bigger is m, the more are the student difficulties during the MM process. Of course, there are also other indications about the student difficulties during the MM process, like the total time spent by students, etc.

An Application of the EMC Model

The model developed in Section 3.3.1 can help the instructor to assess the student MM skills, but also to evaluate his (her) teaching effectiveness. The following classroom experiment illustrates this argument.

Description: Two groups of 20 students were formed from two different departments of the School of Management of the Graduate TEI of Western Greece being at the beginning of their second term of studies. The students were taught the same mathematical topics at their first term of studies by two different instructors, they were examined on subjects of roughly the same difficulty and the mean values of their marks were almost the same (equivalent groups).

Ten MM problems (see below) were handed to the students of both groups for solution in three hours. It was explained to the students that we were interested for all their attempts (wrong and right) to solve the problems. Also, in order to be helped in collecting the appropriate data from their answers, we asked them:

- To mark with an "A'' the problems in which their initial efforts didn't lead to a solution (not including cases of numerical mistakes corrected afterwards).
- To mark with a "B" those problems where they found the solution directly, without making any modifications to their initially constructed models.
- To mark with a "C" the problems in which they have made modifications to their model after realizing that the solution obtained didn't give a reliable prediction of the real situation.

The Problems: *Problem 1:* We want to construct a channel to run water by folding the two edges of an orthogonal metallic leaf having sides of length 20cm and 32 cm, in such a way that they will be perpendicular to the other parts of the leaf. Assuming that the flow of the water is constant, how we can run the maximum possible quantity of the water?

(*Remark:* The correct solution is obtained by folding the edges of the longer side of the leaf)

Problem 2: A car dealer has a mean annual demand of 250 cars, while he receives 30 new cars per month. The annual cost of storing a car is 100 euros and each time he makes a new order he pays an extra amount of 2200 euros for general expenses (transportation, insurance etc). The first cars of a new order arrive at the time when the last car of the previous order has been sold. How many cars must he order in order to achieve the minimum total cost?

Problem 3: An importation company codes the messages for the arrivals of its orders in terms of characters consisting of a combination of the binary elements 0 and 1. If it is known that the arrival of a certain order will take place from 1st until the 16th of March, find the minimal number of the binary elements of each character required for coding this message.

Problem 4: Let us correspond to each letter the number showing its order into the alphabet (A=1, B=2, C=3 etc). Let us correspond also to each word consisting of 4 letters a 2X2 matrix in the obvious way; e.g. the matrix $\begin{bmatrix} 19 & 15 \\ 13 & 5 \end{bmatrix}$ corresponds to the word SOME. Using the matrix E= $\begin{bmatrix} 8 & 5 \\ 11 & 7 \end{bmatrix}$ as an encoding matrix how you could send the message LATE in the form of a camouflaged matrix to a receiver knowing the above process and how he (she) could decode your message?

Problem 5: The demand function $P(Q_d)=25-Q_d^2$ represents the different prices that consumers willing to pay for different quantities Q_d of a good. On the other hand the supply function $P(Q_s)=2Q_s+1$ represents the prices at which different quantities Q_s of the same good will be supplied. If the market's equilibrium occurs at (Q_0, P_0) producers who would supply at lower price than P_0 benefit. Find the total gain to producers'.

Problem 6: A ballot box contains 8 balls numbered from 1 to 8. One makes 3 successive drawings of a lottery, putting back the corresponding ball to the box before the next lottery. Find the probability of getting all the balls that he draws out of the box different.

Problem 7: A box contains 3 white, 4 blue and 6 black balls. If we put out 2 balls, what is the probability of choosing 2 balls of the same colour?

Problem 8: The population of a country is increased proportionally. If the population is doubled in 50 years, in how many years it will be tripled?

Problem 9: A wine producer has a stock of wine greater than 500 and less than 750 kilos. He has calculated that, if he had the double quantity of wine and transferred it to bottles of 12, 25, or 40 kilos, it would be left over 6 kilos each time. Find the quantity of stock.

Problem 10: Among all cylindrical towers having a total surface of 180π m², which one has the maximal volume?

(*Remark*: Some students didn't include to the total surface the one base (ground-floor) and they found another solution, while some others didn't include both bases (roof and ground-floor) and they found no solution, since we cannot construct cylinder with maximal volume from its surrounding surface.)

Calculation of the transition probabilities: For clarifying the way in which we collected the experiment's data, let us consider the answers of the students of the first student group for Problem1:

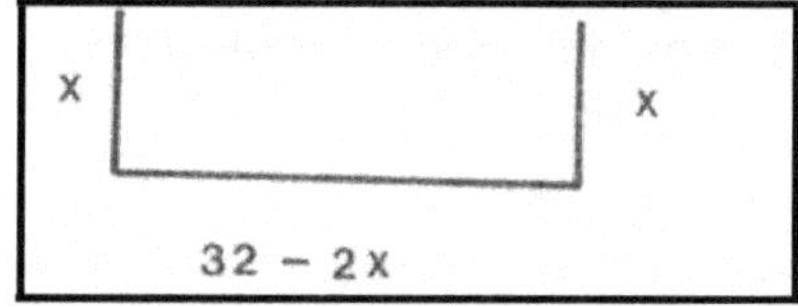

Figure 5: Construction of the channel in Problem 1

Folding the edges of the longer side of the leaf (Figure 8) one has to maximize the function E(x) = (32-2x)x. Thus, taking the derivative E′(x) = 0, he (she) obtains that x = 8cm. Therefore E(8) = 128cm². In this problem 14 students found the correct solution and 10 of them marked it with "B", three marked it with "C" and one student marked it with "A - C". Also four students folded the other side of the leaf and they found E(x) = (20-2x)x, which gives that x = 5cm. In this case E(5) = 50cm², therefore this solution was wrong. The remaining two students didn't succeed to construct a solvable model.

We worked in the same way with the other nine problems and then we calculated the following mean values per problem for the first student group: Correct solutions 11.2; from these solutions 5.7 were marked with B, 2.3 were obtained by making modifications to the initial model only after the state $\mathbf{S_4}$ (marked with C) , and the remaining 3.2 solutions obtained by starting to make modifications to the model just after the first passage from state S_3 (marked with A or A - C). Notice that supplementary modifications became necessary to 1.1 from these 3.2 solutions after state S_4 (marked with A-C). We also found 6.3 wrong solutions and 2.5 failures to construct a solvable model.

Further, the solver "arrivals" and "departures" to each step of the MM process were calculated in a way analogous to that applied in Problem 1 of Section 3.2.3 (Figure 6).

Therefore, since we had 35.4 in total "arrivals" to S_2 and 26.6 "departures" from S_2 to S_3, one finds that p_{23} = 26.6 : 35.4 ≈ 0.751, therefore p_{21} ≈ 0.249. In the same way it turns out that p_{34} = 20.9 : 26.6 ≈ 0.786 , p_{32} ≈ 0.214,
p_{45} ≈ 11.2 : 20.9 ≈ 0.536 and p_{42} ≈ 0.464. Replacing the above values in equations (2) of Section 3.3.1 one finds α_1 = 0.176, α_2 ≈ 0.31, α_3 ≈ 0.233, α_4 ≈ 0.183, and α_5 ≈ 0.098. Replacing these values in equation (3) of Section 3.2.1 one finds that m = 0.902 : 0.098 ≈ 9.204 steps.

The corresponding mean values for the second student group were the following: Correct solutions 8.1; from these solutions 2.1 were marked with B, 0.6 with C, 5.2 with A and 0.2 solutions with A - C. We also found 9.7 wrong solutions and 2.2 failures to construct a solvable model. Working as in the case of the first group one finds for the second student group that p_{23} ≈ 0.688, p_{21} ≈ 0.312, p_{34} ≈ 0.71, p_{32} ≈ 0.29, p_{45} ≈ 0.435 and p_{42} ≈ 0.565, which give that α_1 ≈ 0.18, α_2 ≈ 0.343, α_3 ≈ 0.236, α_4 ≈ 0.168, α_5 ≈ 0.073 and m ≈ 12.699 steps.

Observing the above outcomes one concludes that the step S_2 of mathematization had the greatest "gravity" for both student groups. This was expected, since this step is very crucial for MM requiring a deep abstraction ability from the solver. It becomes also evident that the first group demonstrated a better performance concerning the MM process ($9.204<12.699$). Accordingly, since the groups were equivalent, this provides an indication that the teaching effectiveness of the first group's instructor on subjects involving MM was better than that of the second group's instructor. However, the two groups' performance could be also affected by random factors, like student spiritual condition in the particular moment, silly mistakes, etc.

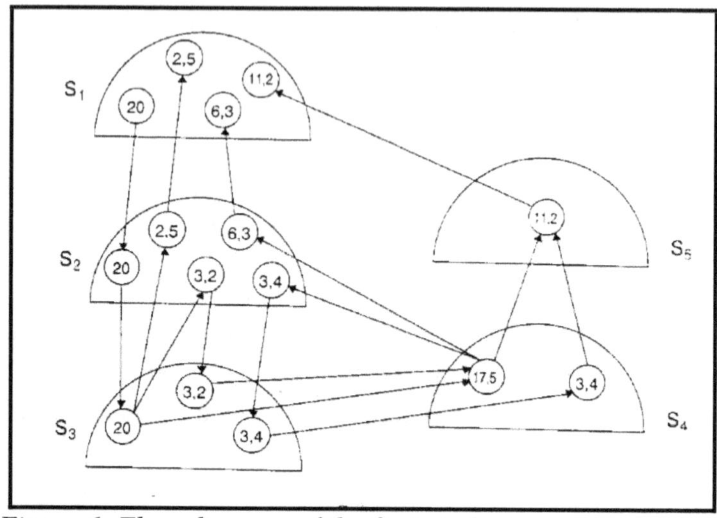

Figure 6: Flow-diagram of the first student group movements

ASSESSING ANALOGICAL REASONING ABILITIES

Analogical Reasoning

Analogies are used for explaining concepts which cannot directly perceived (e.g. electricity in terms of the water flow), in making predictions within domains, in communication and persuasion, etc. ***Analogical Reasoning (AR)*** is a method of processing information that compares the similarities between new and past understood concepts, then using these similarities to gain understanding of the new concept. AR is important in general for creativity and scientific discovery. Within cognitive science mental processes are likened to computer programs (e.g. neural networks) and such analogies serve as mental models to support reasoning in new domains.

Analogical Problem Solving (APS) is the main mechanism of AR: When the solver is not sure of the appropriate procedure to solve a given problem (***target problem***), a good hint would be to look for a similar problem solved in the past (***source problem***) and then try to adapt the solution procedure of this problem for use with the target problem. However this strategy can be difficult to implement in PS, because it requires the solver to attend to information other than the problem to be solved. Thus the solver may come up empty-handed, either because he/she has not solved any similar problems in past, or because he/she fails to realize the relevance of previous problems. But, even if an analogue is retrieved, the solver must know how to use it to determine the solution procedure for the target problem.

Several studies (Getner & Toupin, 1986, Gick & Holyoak, 1983, Holyoak, 1985, Novick, 1988, Voskoglou 2003 , etc.) have provided detailed models for the AR process based on APS, in which factors associated with instances of successful transfer of knowledge are identified. According to these studies the main steps involved in APS include:

- ***Representation*** of the target problem.
- ***Search-retrieval*** of a source problem
- ***Mapping*** of the representations of the target and the source problem.
- ***Adaptation*** of the solution of the source problem for use with the target problem.

More explicitly, before solvers working on a problem they usually construct a representation of it. A good representation must include both the surface and the structural (abstract, solution relevant) features of the problem. The former are mainly determined by what are the quantities involved in the problem and the latter by how these quantities are related to each other. The features included in solver representations of the target problem are used as retrieval cues for a source problem in memory. When the two problems share structural but not surface structures the source is called a ***remote analogue*** of the target problem. Analogical mapping requires aligning the two situations – that is, finding the correspondences between the representations of the target and the source problem – and projecting inferences from the source to the target. Once the common alignment and the candidate inferences have been discovered the analogy is evaluated. The last step involves the adaptation of the solution of the analogous problem for use with the target problem, where the correspondences between objects and relations of the two problems must be used.

The successful completion of the above process is referred as ***positive analogical transfer***. But the search may also yield ***distracting problems*** sharing surface but not structural (solution relevant) common features with the target problem and therefore being only superficially similar to it.

Usually the reason for this is a non satisfactory representation of the target problem, containing only its salient surface features, and the resulting consequences on the retrieval cues available for the search process. When a distracting problem is considered as an analogue of the target, we speak about ***negative analogical transfer***. This happens if a distracting problem is retrieved as a source problem and the solver fails, through the mapping of the representations of the source and target problem, to realize that the source cannot be considered as an analogue to the target. Therefore the process of mapping is very important in APS playing the role of a "*control system*" for the fitness of the source problem.

The MC model

It is assumed here that the steps of the APS process posses the Markov property. This assumption may be considered as a simplification (not far away from the truth) made to the real system that enables the formulation of it to a form ready for mathematical treatment (assumed real system, see Section 1.1). The states of the corresponding MC are: s_1 = representation, s_2 = search-retrieval, s_3 = mapping, s_4 = adaptation of the solution of the source problem and s_5 = solution of the target problem. The starting state is always s_1. When the APS process is completed at s_5, it is assumed that a new problem is given (or appears) for solution and therefore the process restarts from s_1. For describing the AR process this is compatible of considering it a sequence of APS activities

After completing the target problem's representation the solvers proceed from s_1 to s_2. Being at s_2 and facing difficulties in finding a source problem they may return to s_1 looking for more information in problem's representation that possibly has been elapsed at first hand. Then they return to s_2 to continue the APS process.

After the retrieval of a source problem the solvers proceed from s_2 to s_3. If the source is considered to be analogous to the target problem, then they transfer from s_3 to s_4. Otherwise they return to s_2 searching for a new suitable source problem. Notice that solvers who finally fail to retrieve an analogue through the mapping process cannot proceed further. Therefore they return to s_1 waiting for a new problem to be given for solution.

After the adaptation of the solution of the source for use with the target problem the solvers proceed to the final state s_5 of the solution of the target problem. On the contrary, if during the adaptation process they realize that the source is a distracting problem, then they return to s_2 searching for a new source problem. Solvers who finally fail to adapt the solution of the source for use with the target problem they return from s_4 to s_1 waiting for a new problem to be given for solution. According to the above description the *flow-diagram* of the APS process is that shown in Figure 10.

Denote by p_{ij} the transition probabilities from state s_i to s_j, for $i, j = 1, 2, 3, 4, 5$. According to the diagram of Figure 10 the transition matrix of the chain is

$$A = \begin{array}{c} \\ s_1 \\ s_2 \\ s_3 \\ s_4 \\ s_5 \end{array} \begin{array}{c} \begin{array}{ccccc} s_1 & s_2 & s_3 & s_4 & s_5 \end{array} \\ \begin{bmatrix} 0 & 1 & 0 & 0 & 0 \\ p_{21} & 0 & p_{23} & 0 & 0 \\ p_{31} & p_{32} & 0 & p_{34} & 0 \\ p_{41} & p_{42} & 0 & 0 & p_{45} \\ 1 & 0 & 0 & 0 & 0 \end{bmatrix} \end{array}$$

with $p_{21}+ p_{23} = p_{31}+p_{32}+p_{34} = p_{41}+p_{45} =1$ (1).

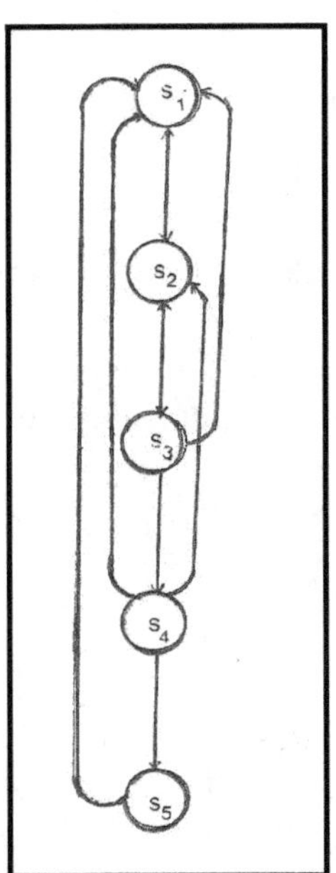

Figure 7: The flow-diagram of the APS process

Obviously the chain is an EMC, therefore P = PA (2), where P = [p_1 p_2 p_3 p_4 p_5] is the limiting probability vector, with. $p_1+p_2+p_3+p_4+p_5=1$ (3).

From relation (2) one gets easily the following equations:
$p_1=p_2p_{21}+p_3p_{31}+p_4p_{41}+p_5$, $p_2=p_1+p_3p_{32}+p_4p_{42}$, $p_3=p_2p_{23}$, $p_4=p_3p_{34}$, $p_5=p_4p_{45}$

Adding the first four of the above equations and using relation (5) one finds the fifth equation, which therefore is equivalent with the others.

Solving the linear system of the first four equations and of equation (3) by the Cramer's rule one finds that $p_1 = \frac{(p_{31}-1)(p_{41}-1)+p_{23}p_{32}(p_{41}-1)-p_{23}p_{42}p_{34}}{D}$,

$p_2=\frac{(p_{31}-1)(p_{41}-1)}{D}$, $p_3=\frac{-p_{23}(p_{41}-1)}{D}$, $p_4=\frac{p_{23}p_{34}}{D}$,

$p_5=\frac{-p_{23}p_{34}(p_{41}-1)-p_{23}p_{34}p_{42}}{D}$ (4).

Here $D=(2+p_{23})(p_{31}-1)(p_{41}-1)+p_{23}(2p_{32}-1)(p_{41}-1)+p_{23}p_{34}(1-2p_{42})$ is the determinant of the system.

Further, the fraction $\frac{p_i}{p_j}$ calculates the mean number of times in state s_i between two successive occurrences of s_j . Therefore, since the process restarts again from s_1 after the completion of state s_5, the mean number of times between two successive occurrences of s_5 is given by

$$m=\sum_{i=1}^{4}\frac{p_i}{p_5}=\frac{1-p_5}{p_5} \quad (5).$$

The value of m is an indicator of the solver difficulties during the APS process, another indicator is the time spent for the solution of each problem. However, assuming that the time available for the solution of each problem is prefixed, the value of m can be considered as a measure for solver difficulties during the APS process. The bigger is m the more the solver difficulties during the AR process.

A classroom experiment

For illustrating the use of the above model in practice we performed the following experiment (Voskoglou, 2013b) in which the subjects were students of the Graduate Technological Educational Institute of Patras, Greece, being at their second term of studies. We formed two groups, with 20 students of the School of Management and Economics in the first and 20 students of the School of Technological Applications (prospective engineers) in the second group.

Three mathematical problems were given for solution to both groups on common topics of the student first term course in mathematics. In each case and before receiving the target problem the students received two other problems together with their solution procedures. They read each problem and its solution procedure and then solved the problem themselves using the given procedure. Subjects were allowed 10 minutes for each problem and they were not given the other problem until after 10 minutes had elapsed. The first of these problems was a remote analogue to the target problem, while the other was a distracting problem. Next the target problem was given and was asked from the students to try to solve it by adapting the solution of one of the previous problems (time allowed 20 minutes). Our instructions stressed the importance of showing all of one's work on paper and emphasized that we were interested in both correct and incorrect solution attempts. The problems given to students are presented in the next Section.

Examining student papers after the end of the experiment we calculated the following means:

- 4.2 students from the first group faced difficulties in retrieving a source problem, but they came through after looking back to their representations of the target problem (5.1 students from the second group).

- 15.1 students from the first group considered through the mapping the collected source as an analogue to the target problem, while the rest of them (4.9 students) searched for a new source. Finally 3.7 of them considered the new source as an analogue to the target problem, while the rest of them (1.2 students) failed to retrieve an analogue through the mapping process (14.8 and 1.6 students from the second group respectively).

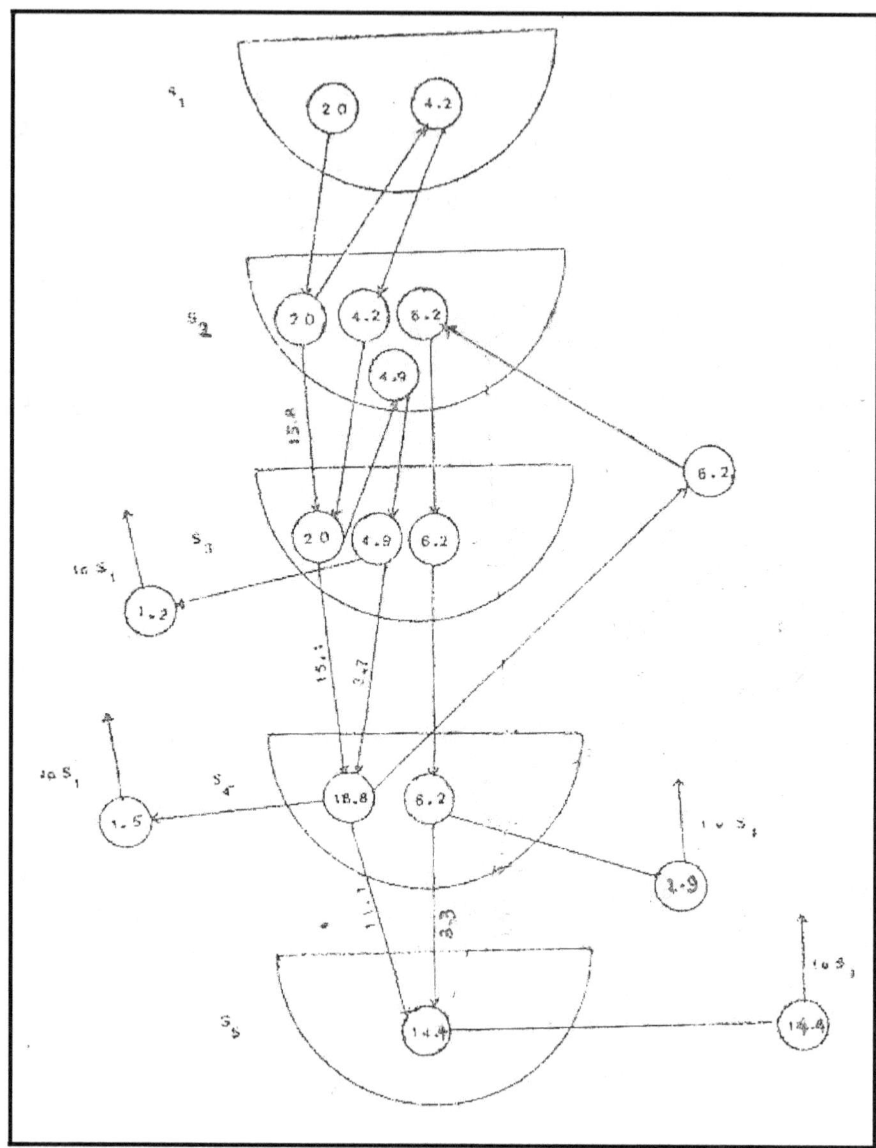

Figure 8: The movements of the students of the first group

- Thus 15.1+3.7=18.8 students from the first group proceeded to the step of adaptation (14.8+1.6=16.4 students from the second group). From these students 11.1 adapted successfully the solution of the analogue for use with the target problem, while 1.5 students failed to do so.(12.1 and 1.5 students from the second group respectively). The rest of them (6.2 students from the first and 2.8 from the second group) returned to s_2 to retrieve a new source and through s_3 they came back to s_4.

- Finally 3.3 from these 6.2 students of the first group adapted successfully the solution of the analogue for use with the target problem and 2.9 failed to do so (1.6 and 1.2 students from the second group respectively). Thus 11.1+3.3=14.4 students from the first and 12.1+1.6=13.7 students from the second group solved the target problems.

The movements of the students of the first group are shown in Figure 8. We observe that we have 35.3 in total “arrivals” to s_2 and 31.1 “departures”

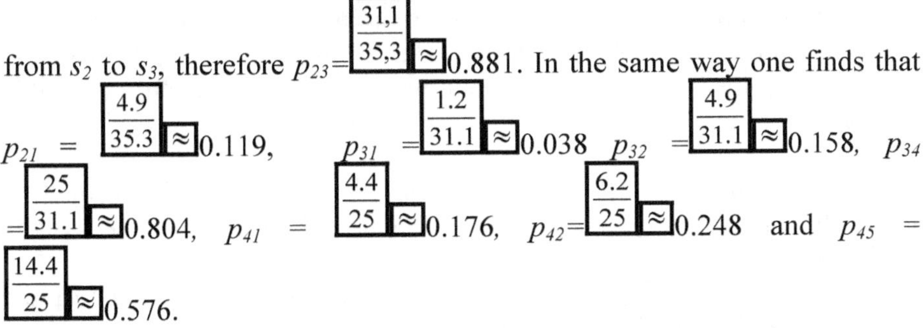

from s_2 to s_3, therefore $p_{23}=\frac{31,1}{35,3}\approx 0.881$. In the same way one finds that $p_{21}=\frac{4.9}{35.3}\approx 0.119$, $p_{31}=\frac{1.2}{31.1}\approx 0.038$ $p_{32}=\frac{4.9}{31.1}\approx 0.158$, $p_{34}=\frac{25}{31.1}\approx 0.804$, $p_{41}=\frac{4.4}{25}\approx 0.176$, $p_{42}=\frac{6.2}{25}\approx 0.248$ and $p_{45}=\frac{14.4}{25}\approx 0.576$.

Replacing the values of the transition probabilities in the formulas (4) and (5) of Section 3.4.2 one finds that the limiting probability vector for the first group is
$P\approx$[0.157 0.259 0.231 0.232 0.121] and that $m\approx 7.264$ times.

Making the analogous calculations for the second group one finds that $P\approx$[0.154 0.26 0.237 0.23 0.119] and $m\approx 7.404$ times. The elements of P give the several probabilities about the “behavior” of each group during the AR process. Also, since 7,264<7,404, the performance of the first group was slightly better.

According to the design of our experiment students had to choose the source problem between two given problems: A remote analogue to the target and a distracting problem. However, often things are not so simple. In fact, the individuals have usually to search in their memories to retrieve the source among several past problems sharing common surface and/or structural characteristics with the target. We could of course add in our experiment one or more problems among the candidate source problems. Nevertheless, this manipulation would make the calculation of the transition probabilities between states of the chain more complicated, because the students movements would be extended to other directions as well.

The problems of the experiment

CASE 1

Target problem: A box contains 8 balls numbered from 1 to 8. One makes three successive drawings, putting back the corresponding ball to the box before the next drawing. Find the probability of getting all the balls drawing out of the box different to each other.

The probability is equal to the quotient of the total number of the ordered samples of 3 objects from 8 (favourable outcomes) to the total number of the corresponding samples with replacement (possible outcomes).

Remote analogue: How many numbers of 2 digits can be formed by using the digits from 1 to 6 and how many of them have their digits different?
Solution procedure given to the students: Find the total number of the ordered samples of 2 objects from 6 with and without replacement respectively.

Distracting problem: A box contains 3 white, 4 blue and 6 black balls. If we draw out 2 balls, what is the probability to be of the same colour?
Solution procedure given to the students: The number of all favourable outcomes is equal to the sum of the total number of combinations of 3, 4 and 6 objects taken 2 at each time respectively, while the number of all possible outcomes is equal to the total number of combinations of 13 objects taken 2 at each time.

CASE 2

Target problem: Consider the matrices:

$$A = \begin{bmatrix} 1 & -á & -á \\ 0 & 1 & -á \\ 0 & 0 & 1 \end{bmatrix} \quad \kappa\alpha\iota \quad B = \begin{bmatrix} 0 & -á & -á \\ 0 & 0 & -á \\ 0 & 0 & 0 \end{bmatrix}.$$

Prove that $A^n = A + (n-1)(B + \frac{n}{2}B)$, for every positive integer n. -

Since A=I+B, where I stands for the unitary 3X3 matrix, and $B^3 = 0$, is $A^n = (I+B)^n = I + nB + \frac{n(n-1)}{2}B^2 = A + (n-1)B + \frac{n(n-1)}{2}B = A + (n-1)(B + \frac{n}{2}B)$.

Remote analogue: Let α be a nonzero real number. Prove that $\alpha^n = \sum_{i=0}^{n}\binom{n}{i}(a-1)^n$, for all positive integers n.

Solution procedure given to the students: Write α = 1+(α-1) and apply the Newton's formula $(x+b)^n = \sum_{i=0}^{n}\binom{n}{i}x^{n-i}b^i$, setting x=1 and b=α-1.

Distracting problem: If A and B are as in the target problem, calculate $(A+B)^2$. -

The students were asked to operate the corresponding calculations.

CASE 3

Target problem: The price of sale of a good depends upon its total demand Q and it is given by $P(Q) = \frac{1}{2}Q - 50$, while the cost of production of the good is given by $C(Q) = \frac{1}{4}Q^2 + 35Q + 25$. Find the quantity Q of the good's total demand maximizing the profit from sale.-

The revenue from sale is equal to P(Q)Q and therefore the profit from sale is given by K(Q) = P(Q)Q-C(Q). The maximum of function K(Q) is calculated by using the derivatives.

Remote analogue: A car is entering in a road having initial speed 50 Km/h, which is changed according to the relation $U(t) = 3t^2 - 12t + 50$, where t represents the time (in minutes) during which the car is moving on this road. Find the minimal speed of the car on this road.-

The students were asked to calculate the minimum of the function U(t) using the derivatives.

Distracting problem: The price of sale of a good depends upon its total demand Q and it is given by $P(Q)=25-Q^2$. The price is finally fixed to 9 monetary units and therefore the consumers who would be planning to pay more than this price benefit. Find the total benefit to consumers (Dowling 1980, paragraph 17.7: *Consumer's surplus*).

Solution procedure given to the students: For P=9 and since Q ≥ 0, it turns out that Q=4.metric units. Drawing the graph of the function P(Q) (parabola) it is easy to observe that the total benefit to consumers is equal to

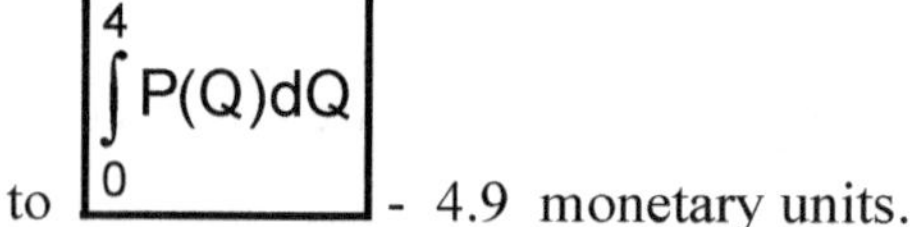

- 4.9 monetary units.

EVALUATING THE EFFECTIVENESS OF CASE-BASED REASONING SYSTEMS

Case-Based Reasoning

As we have already seen before the importance of AR in human thinking has been recognized years ago. However, it is the ***Case-Based Reasoning (CBR)*** approach for PS and learning that has got a lot of attention over the last few years, because as an intelligent-systems method enables information managers to increase efficiency and reduce cost of many human activities by substantially automating processes, such as diagnosis, scheduling and design (Voskoglou & Salem, 2014: Section 3). Note that the term AR is sometimes used as a synonymous of the typical CBR approach (Veloso & Carbonel, 1993). Nevertheless, it is often used also to characterize methods, that solve new problems based on past cases of *different domains* (Hall, 1989, Kedar-Cabelli, 1988) while typical CBR methods focus on single-domain cases, i.e. on a form of *intra-domain analogy*.

CBR is often used where experts find it hard to articulate their thought processes when solving problems. This is because knowledge acquisition for a classical knowledge-based system would be extremely difficult in such domains, and is likely to produce incomplete or inaccurate results. When using CBR the need for knowledge acquisition can be limited to establishing how to characterize *cases*, i.e. the analogous problem situations. A ***case-library*** can be a powerful corporate resource allowing everyone in an organization to tap in the corporate library, when handling a new problem. A ***CBR system,*** usually designed and functioning with the help of computers, allows the case-library to be developed incrementally, while its maintenance is relatively easy and can be carried out by domain experts.

There are two styles of CBR, the ***PS style*** and the ***interpretive style***. The PS style can support a variety of tasks including planning, diagnosis and design, e.g. in Medicine (Silvana, Pedro & Steen, 2001) Industry (Hinkle & Toomey, 1995) and Robotics (Hans-Dieter, Salem & El Bagoury, 2007). The interpretive style is useful for classification, evaluation or justification of a solution, argumentation and for the projection of effects of a decision. Lawyers and managers making strategic decisions use the interpretive style (Rissland & Danials, 1995, Salem & Baeshen, 1999).

CBR has been formalized for purposes of computer and human reasoning as a four steps process, often referred as the “four R’s”. These steps involve:

- ***R_1: Retrieve*** the most similar to the new problem past case.
- ***R_2: Reuse*** the information and knowledge of the retrieved case for the solution of the new problem.
- ***R_3: Revise*** the proposed solution.
- ***R_4: Retain*** the part of this experience likely to be useful for future problem solving.

The first three of the above steps are not linear, characterized by a backward - forward flow among them. A simplified flow - chart of the CBR process, adequate for the purposes of the present paper, is presented in Figure 9 below:

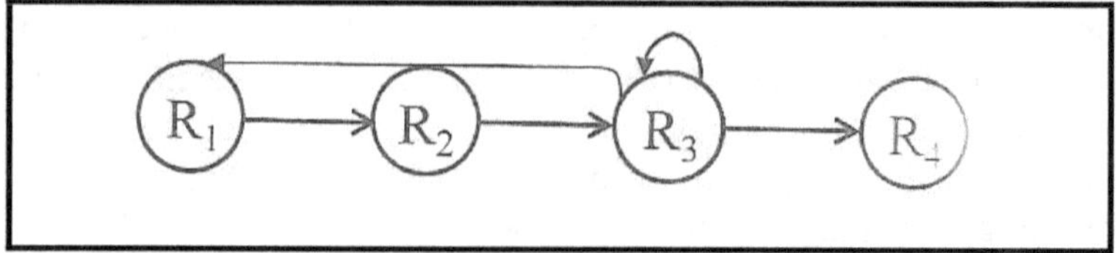

Figure 9: A simplified flow-chart of the CBR process

More details about the CBR methodology, history and applications can be found in Voskoglou & Salem, 2014, Voskoglou, 2008 and in the relevant references given in the previous papers. A detailed functional diagram illustrating the four steps of the CBR process is also available in Voskoglou & Salem, 2014, Figure 1.

As a general PS methodology intended to cover a wide range of real-world applications, CBR must face the challenge to deal with uncertain, incomplete and vague information. Correspondingly recent years have witnessed an increased interest in formalizing parts of the CBR methodology within frameworks of reasoning under uncertainty, and in building hybrid approaches by combining CBR with methods of uncertain and approximate reasoning.

In an earlier work the present author has developed a stochastic model for the CBR process by introducing a finite Markov Chain on it main steps (Voskoglou, 2013a). Here this model will be used to develop a method for evaluating the effectiveness of CBR systems.

The MC model for CBR

Assuming that the CBR process has the Markov property, we introduce a finite MC having as states the four steps of the CBR process described in Section 3.5.1. Denote by p_{ij} the transition probability from state R_i to R_j, for i, j = 1, 2, 3, 4. According to the flow-diagram of Figure 12 one finds that the transition matrix of the MC is

$$A = \begin{array}{c} \\ R_1 \\ R_2 \\ R_3 \\ R_4 \end{array} \begin{array}{c} \begin{array}{cccc} R_1 & R_2 & R_3 & R_4 \end{array} \\ \begin{bmatrix} 0 & 1 & 0 & 0 \\ 0 & 0 & 1 & 0 \\ p_{31} & 0 & p_{33} & p_{34} \\ 0 & 0 & 0 & 1 \end{bmatrix} \end{array},$$

with $p_{31}+p_{33}+p_{34} = 1$.

Let us denote by $\varphi_0, \varphi_1, \varphi_2, \ldots\ldots$ .. the successive phases MC , and let $P_i=[p_1^{(i)}\ p_2^{(i)}\ p_3^{(i)}\ p_4^{(i)}]$ be the probability vector of the chain at phase φ_i, where $\sum_{j=1}^{4} p_j^{(i)} = 1$.

From the transition matrix A and the flow diagram of Figure 12 one obtain the ***tree of correspondence*** among the phases and states of the MC shown in Figure10.

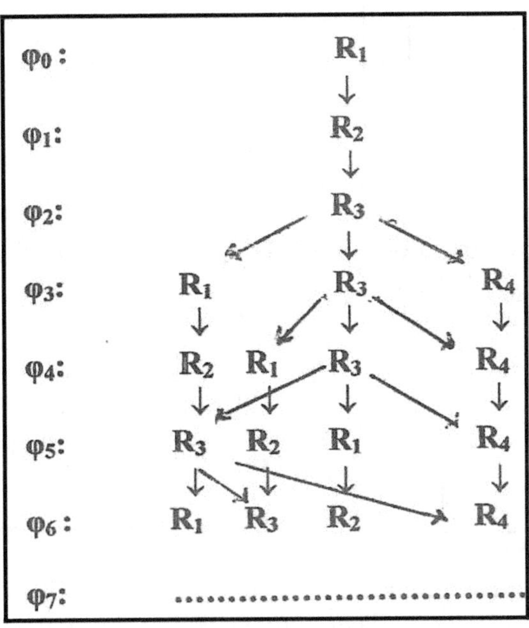

Figure 10: Tree of correspondence of the MC

From the above tree it becomes evident that

$P_0 = [1\ 0\ 0\ 0]$, $P_1 = [0\ 1\ 0\ 0]$, $P_2 = [0\ 0\ 1\ 0]$, and $P_3 = [p_{31}\ 0\ p_{33}\ p_{34}]$.

Further, from $P_{i+1} = P_iA$, $i = 0, 1, 2, \ldots$, one finds that

$P_4 = P_3A = [p_{33}p_{31}\ \ p_{31}\ \ p_{33}^2\ \ p_{34}(p_{33}+1)]$

$P_5 = P_4A = [p_{33}^2p_{31}\ p_{33}p_{31}\ p_{31}+p_{33}^3\ p_{34}(p_{33}^2+p_{33}+1)]$ and so on.

Observe now that, when the chain reaches the state R_4, it is impossible to leave it, because the solution process of the new problem via the CBR approach finishes there. Thus we have an *AMC* with R_4 its unique absorbing state. The canonical form A* of the transition matrix is given by

$$A^* = \begin{array}{c} \\ R_4 \\ - \\ R_1 \\ R_2 \\ R_3 \end{array}\begin{array}{c} R_4 \quad\quad R_1 \quad R_2 \quad R_3 \\ \left[\begin{array}{c|ccc} 1 & 0 & 0 & 0 \\ \hline 0 & 0 & 1 & 0 \\ 0 & 0 & 0 & 1 \\ p_{34} & p_{31} & 0 & p_{33} \end{array}\right] \end{array}.$$

In the above partition of A* the bottom-right hand matrix Q is the transition matrix between the non absorbing states, therefore the fundamental matrix of the chain is given by

$$N = (I_3 - Q)^{-1} = \frac{1}{1-p_{31}-p_{33}} \begin{bmatrix} 1-p_{33} & 1-p_{33} & 1 \\ -p_{31} & 1-p_{33} & 1 \\ p_{31} & p_{31} & 1 \end{bmatrix} = [n_{ij}]$$

The entry n_{ij} of N gives the mean number of times in state R_j when the chain is started in state R_i. Therefore, since the starting state is always R_1, the sum

$$t = n_{11}+n_{12}+n_{13} = \frac{3-2p_{33}}{1-p_{31}-p_{33}} \quad (1)$$

gives the mean number of phases of the chain before absorption. Consequently, the mean number of steps for the completion of the CBR process is equal to t+1.

Obviously, the bigger is the value of t, the greater is the difficulty encountered for the solution of the given problem via the CBR process. The ideal case is when the CBR process is completed straightforwardly, i.e. without "backwards" from R_3 to R_1, or "stays" to R_3 (see Figure 12). In this case we have that $p_{31}=p_{33}=0$ and $p_{34}=1$, therefore t=3. Thus in general we have that $t \geq 3$.

The following simple example illustrates the above results.

Example 1

A physician takes into account the diagnosis and treatment of a previous patient having similar symptoms in order to determine the disease and treatment for the patient in front of him. Obviously the physician is using CBR. If the initial treatment fails to improve the health of the patient, then the physician either revises the treatment (stay to R_3 for two successive phases), or, in more difficult cases, gets a reminding of a previous similar failure and uses the failure case to improve its understanding of the present failure and correct it (transfer from R_3 to R_1). The process is completed, when the physician succeeds to cure the patient.

The statistical data show that the probabilities of a straightforward cure of the patient and of each of the above two reactions of the physician in case of failure are equal to each other. This means that $p_{13} = p_{33} = p_{34} = \frac{1}{3}$ and therefore formula (1) gives that t=7, i.e. the mean number of steps for the cure of the patient is 8.

Further, one finds that $P_3 = [\frac{1}{3}\ \frac{1}{3}\ \square\ \frac{1}{3}]$, $P_4 = [\frac{1}{9}\ \frac{1}{3}\ \frac{1}{9}\ \frac{4}{9}]$, $P_5 = [\frac{1}{27}\ \frac{1}{9}\ \frac{4}{9}\ \frac{13}{27}]$ and so on.

Observing for example the probability vector P_5 one finds that the probability for the CBR process to be at the step of revision (R_3) in the 6th phase after the beginning of the patient's therapy is $\frac{4}{9}$, or approximately 44.44%, the corresponding probability to be at the step of retaining the acquired experience (R_4) is $\frac{13}{27}$, or approximately 48.15%, etc.

Effectiveness of CBR systems

The challenge in CBR is to come up with methods that are suited for problem-solving and learning in particular subject domains and for particular application environments. In line with the process model described in Section 3.5.2, core problems addressed by CBR research can be grouped into five areas involving representation of cases and methods for retrieval, reuse, revision and retaining the acquired experience. A CBR system should support the problems appearing in the above five areas. A good system should support a variety of retrieval mechanisms and allow them to be mixed when necessary. In addition, the system should be able to handle large case libraries with the retrieval time increasing linearly with the number of cases.

Let us consider now a CBR system including a library of *n* recorded past cases and let t_i be the mean number of steps for the completion of the CBR process for case c_i, i = 1, 2,..., *n*, as it is calculated by formula (10) of Section 3.5.2. Each t_i could be stored in the system's library together with the corresponding case c_i. We define then the system's ***effectiveness***, say t, to be the mean value of the t_i's of its stored cases, i.e. we have that

$$t = \frac{\sum_{i=1}^{n} t_i}{n} .$$

The more problems solved by the given system, the bigger becomes the number *n* of the stored cases in the system's library and therefore the value of t is changing. As *n* increases it is normally expected that t will decrease, because the values of the t_i's of the new stored cases would be decreasing. In fact, the bigger is *n*, the better would be the chance of a new case to "fit" well (i.e. to have minor differences) with a known past case, and therefore the less would be the difficulty of solving the corresponding problem via the CBR process. Thus we could say that a CBR system *behaves well* if, when *n* tends to infinity, then its efficiency t tends to its maximal value, which, as we have seen in Section 3.5.2, is equal to 3.

The following example illustrates the use of the above definition in practice.

Example 2

Consider a CBR system has been designed in terms of the Schank's model *of* ***dynamic memory*** (Schank, 1982). The basic idea of this model is to organize specific cases, which share similar properties, under a more general structure called a ***generalized episode (GE)***. During the storing of a new case, when the features of it match the features of an existing past case not belonging with other past cases to an already existing GE, a new GE is created. Hence the memory structure of the system is in fact dynamic, in the sense that similar parts of two case descriptions are dynamically generalized in to a new GE and the cases are indexed under this GE by their different features.

In order to calculate the effectiveness of a system of this type one needs first to calculate the partial effectiveness of each GE contained in it. For example, assume that the given system contains a GE including three cases, say c_1, c_2 and c_3. Assume further that c_1 corresponds to a straightforward successful application of the CBR process, that c_2 is the case described in the example of section 3.5.3 and that c_3 includes one "return" from R_3 to R_1 and two "stays" to R_3. Then $t_1=3$ and $t_2=7$, while for the calculation of t_3 observe with the help of Figure 12 that $p_{31} = p_{34}$

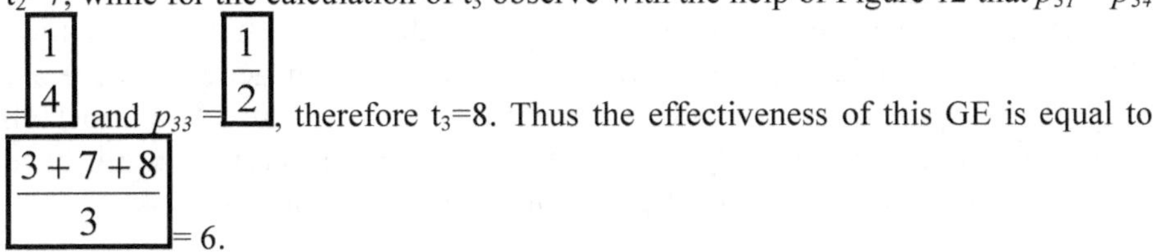

$= \frac{1}{4}$ and $p_{33} = \frac{1}{2}$, therefore $t_3=8$. Thus the effectiveness of this GE is equal to $\frac{3+7+8}{3} = 6$.

Note that a complex GE may contain some more specific GE's (e.g. see Figure 3 in page 12 of Aamodt & Plaza, 2004). In this case we calculate the effectiveness of the complex GE by considering all its cases only once, regardless if they belong to one or more of the specific GE's contained in it. Finally, the effectiveness of the system is the mean value of the effectiveness of each one of its GE's.

An alternative approach for the representation of cases in a CBR system is the ***category and exemplar model*** applied first to the PROTOS system (Porter & Bareiss, 1986). In this model the case memory is embedded in a network of categories, cases and index pointers. A new case is stored in a category by searching for a matching case and by establishing the appropriate feature indices. The process of calculating the effectiveness of a system of this type is analogous to the process applied for systems with dynamic memory, the only difference being that one has to work with categories instead of GE's. In a similar way one may calculate the effectiveness of systems corresponding to other case memory models including Rissland's and Ashley's HYPO system in which cases are grouped under a set of domain-specific dimensions (Rissland, 1983), the MBR model of Stanfill & Waltz (1988), designed for parallel computation rather than knowledge-based matching, etc.

OTHER APPLICATIONS

In Voskoglou (1996) an AMC was utilized for the description of the process of learning a subject matter in the classroom. Also, in Voskoglou (1994) an AMC was introduced on the major steps of the modelling process (in general, not MM only) of a real system and some useful conclusions were obtained. Note that other researchers have also applied frequently in the past principles of MC theory on learning contexts. For example, Suppes and Atkison (1960) introduced MC learning models for multi-person interaction, Perdikaris studied the decision-making process during teaching (Perdikaris, 1992) and he also developed a method of distinguishing between different types of students' geometric reasoning through the van Hiele levels (Perdikaris, 1994), etc. In general, there are very many applications of MC's reported in the literature to almost all sectors of the human activity. However, a complete reference to all, or at least to the most important, of them is out of the scope of this book.

CONCLUSION

Five MC models developed in this Chapter for a mathematical description of the DM, PS, MM, AR and CBR processes. From the discussion performed it became evident that all these models are self restricted to describe the human (or machine) ***assumed behavior*** during the corresponding process. In fact, although the mathematical development of these models was proved to be smart and straightforward, the calculation of the transition probabilities involved in the examples was based on how a human could act and not how he (she) really acts in practice. This is due to the fact that Probability Theory, which is based on principles of the Aristotle's bi-valued logic, apart from making forecasts about possible actions, it has not the capability to describe precisely the ***"acting in the moment"*** human behavior. Therefore, although the stochastic models in general and the MC models in particular are undoubtedly very useful for dealing with real situations characterized by uncertainty, there is a need to find a way to complete them in order to be able to describe the human behavior more precisely. This is a role that could be played by the Zadeh's ***Fuzzy Logic***, principles and applications of which will be studied in the second part of this book.

REFERENCES

Aamodt, A. & Plaza, E. (1994), Case-Based Reasoning:: Foundational Issues, Methodological Variations, and System Approaches, *A. I. Communications*, 7(1), 39-52.

Berger J. O. (1980), *Statistical Decision Theory: Foundations, Concepts and Methods*, Springer-Verlag, New York.

Carlson, M.P. & Bloom, I. (2005), The cyclic nature of problem solving: An emergent multidimensional problem-solving framework, *Educational studies in Mathematics*, 58, 45-75.

Genter, D. and Toupin, C. (1986), Systematicity and surface similarity in development of analogy, *Cognitive Science*, 10, 277-300.

Gick, M.L. & Holyoak, K.J. (1983), Schema induction and analogical transfer, *Cognitive Psychology*, 15, 1-38.

Hall, R. P. (1989), Computational approaches to analogical reasoning: A comparative analysis, *Artificial Intelligence*, 39 (1), 39-120.

Hans-Dieter, Salem A. B. & El Bagoury, B. M. (2007), Ideas of Case-Based Reasoning for Keyframe Technique, *Proceedings of the 16th International Workshop on the Concurrency Specification and Programming*, Logow, Warsawa, Poland, pp. 100-106.

Hinkle, D. & Toomey, C. (1995), Applying Case-Based Reasoning to Manufacturing, *AI Magazine*, 65-73.

Holyoak, K. J. (1985), The pragmatics of analogical transfer, in: G. H. Bower (Ed.), *The psychology of learning and motivation,* Vol. 19, Academic Press, New York, pp. 59-87.

Kedar-Cabelli, S. (1988), Analogy – from a unified perspective, in Helman, D. H. (Ed.), *Analogical Reasoning*, 65-103, Kluwer Academic.

Novick, L. R. (1988), Analogical transfer, problem similarity and expertise, *Journal of Educational Psychology: Learning, Memory and Cognition*, 14, 510-520.

Perdikaris S. C. (1992), A Markov chain model in teachers' decision making, *International Journal of Mathematical Education in Science and. Technology*, 23, 473-477.

Perdikaris S. C. (1994), Markov chains and van Hiele levels: a method of distinguishing different types of students' geometric reasoning processes, *International Journal of Mathematical Education in Science and. Technology*, 25, 585-589.

Polya, G. (1945), *How to solve it,* Princeton Univ. Press, Princeton.

Polya, G. (1954), *Mathematics and Plausible Reasoning* (2 Volumes) , Princeton Univ. Press, Princeton.

Polya G. (1962/65), Mathematical Discovery (2 Volumes), J.Wilet & Sons, New York.

Polya, G. (1963), On learning, teaching and learning teaching, *American Mathematical Monthly,* 70, 605-619.

Polya, G. (1973), *How I solve it: A new aspect of mathematical method,* New Jersey: Princeton University Press.

Porter, B. & Bareiss, B. (1986), PROTOS: An experiment in knowledge acquisition for heuristic classification tasks, *Proceedings of the 1st International Meeting on Advances in Learning*, 159-174, Les Arcs, France.

Rissland, E. (1983), Examples in legal reasoning: Legal hypotheticals. In *Proceedings of the 8th International Joint Conference on Artificial Intelligence* (IJCAI), Karlsruhe

Rissland, E. L. & Danials, J. J. (1995), A Hybrid CBR-IR Approach to Legal Information Retrieval, *Proceedings of the Fifth International Conference on Artificial Intelligence and Law*, pp. 52-61, College Park, MD.

Salem, A-B. M. & Baeshen, N. (1999), Artificial Intelligence Methodologies for Developing Decision Aiding Systems, *Proceedings of 5th International Conference, Integrating Technology and Human Decisions: Global Bridges into the 21st Century*, Decision Sciences Institute, Athens, Greece, pp.168-170.

Schank, R. (1982), *Dynamic memory: A theory of reminding and learning in computers and people,* Cambridge Univ. Press.

Schoenfeld, A. (1980), Teaching Problem Solving skills, *Amer. Math. Monthly*, 87, 794-805.

Schoenfeld, A. (2010), *How we think: A theory of goal-oriented decision making and its educational applications;* Routledge: New York.

Silvana, Q., Pedro, B. & Steen, A. (2001), *Proceedings of 8th Conference on Artificial Intelligence in Medicine in Europe*, Springer, Cascais, Portugal.

Stanfill, C. & Waltz, D. (1988), The memory-based reasoning paradigm, *Case-based reasoning: Proceedings from a workshop*, Morgan Kaufmann, Clearwater Beach, Florida, pp.414-424

Suppes, P. & Atkinson, R. C. (1960), *Markov Learning Models for Multiperson Interactions*, Stanford University Press, Stanford-California, USA.

Veloso, M. M. & Carbonell J. (1993), Derivational analogy in PRODIGY, *Machine Learning*, 10(3), 249-278.

Von Glaserseld, E. (1987), Learning as a Constructive Activity, In C. Janvier (Ed), *Problems of representation in the teaching and learning of mathematics*, Lawrence Erlbaum, Hillsdale, N. J.

Voskoglou, M. Gr. & Perdikaris, S. C. (1991), A Markov chain model in problem- solving, *International Journal of Mathematical Education in Science and. Technology*, 22, 909-914.

Voskoglou, M. Gr. & Perdikaris, S. C. (1993), Measuring problem solving skills, *International Journal of Mathematical Education in Science and. Technology*,, 24, 443-447.

Voskoglou, M. Gr. (1994), An application of Markov chains to the process of modelling, *International Journal of Mathematical Education in Science and. Technology*, 25, 475-480.

Voskoglou, M. Gr. (1995), Measuring Mathematical Model Building Abilities, *International Journal of Mathematical Education in Science and. Technology*, 26, 29-35.

Voskoglou, M. Gr. (1996), The use of Markov Chains to describe the process of learning, *Theta: A Journal of Mathematics* (Manchester Metropolitan University), 10(1), 36-40.

Voskoglou, M. Gr. (2003), Analogical problem solving and transfer, in Gagatsis, A. & Papastavridis, S. (Eds.), *Mathematics in the Modern World: Didactics, Life and Society*, Hellenic Mathematical Society, Athens, , pp. 295-303.

Voskoglou, M. Gr. (2007), A stochastic model for the modelling process. In: *Mathematical Modelling: Education, Engineering and Economics* (ICTMA 12), Chaines, Chr., Galbraith, P., Blum W. and Khan, S. (Eds), 149-157, Horwood Publishing, Chichester, England.

Voskoglou, M. Gr. (2008), Case-Based Reasoning: A Recent Theory for Problem-Solving and Learning in Computers and People, *Communications in Computer and Information Science* (Springer), 18, 314-319.

Voskoglou, M. Gr. (2011), Problem-Solving from Polya to Nowadays: A review and Future Perspectives, in Nata, R. (Ed.), Progress in Education, Vol. 22, Chapter 4, 65-82, Nova Publishers, New York.

Voskoglou, M. Gr. (2013a), Case-Based Reasoning in Computers and Human Cognition: A Mathematical Framework, *International Journal of Machine Intelligence and Sensory Signal Processing,* 1, 3-22.

Voskoglou, M. Gr. (2013b), Probability and Fuzzy Logic in Analogical Reasoning, Journal of Physical Sciences (Vidyasagar University, India), 17, 11-31.

Voskoglou, M. Gr. (2014), Probability and Fuzziness in Decision Making, *Egyptian Computer Science Journal,* 38(3), 86-99.

Voskoglou, M. Gr. & Salem A.-B. (2014), Analogy-Based and Case-Based Reasoning: Two Sides of the Same Coin, *International Journal of Applications of Fuzzy Sets and Artificial Intelligence*, 4, 7-18.

Voskoglou, M. Gr. (2015), Mathematical Modelling as a Teaching Method of Mathematics, *Journal for Research in Innovative Teaching* (National University, CA, USA), 8(1), 35-50.

Voskoglou, M. Gr. (2016), Problem-Solving in the Forthcoming Era of the Third Industrial Revolution, *International Journal of Psychology Research*, 10(4), 361-380.

M. Voskoglou: Finite Markov Chain and Fuzzy Logic Assessment Models

PART II

FUZZY SETS AND SYSTEMS

CHAPTER 4
Fuzzy Sets and Logic

ABSTRACT

In this Chapter, after a short introduction to the basics of Fuzzy Sets and Logic needed for the rest of the book, the process of Decision-Making in a fuzzy environment is studied and a general method is developed for evaluating a system's fuzzy data. The usefulness of this method is illustrated through applications to market's research and to the process of learning a subject matter in the classroom. Further, its general character is emphasized by giving references for its applications to other areas of human cognition developed in earlier author's works.

FUZZY SETS: DEFINITION AND EXAMPLES

The fuzzy sets theory was created in response of expressing mathematically real world situations in which definitions have not clear boundaries; like *"the high mountains of a country"*, *"the young people of a city"*, *"the good players of a team"*, etc. It is recalled that the notion of a fuzzy set was introduced by Zadeh (1965) as follows:

A ***Fuzzy Set (FS)*** *A on the* universal set *U* of the discourse (or a fuzzy subset *A* of *U*) is a set of ordered pairs of the form $A = \{(x, m_A(x)): x \in U\}$, defined in terms of a ***membership function*** $m_A : U \rightarrow [0,1]$ that assigns to each element of *U* a real value from the interval [0,1].

The value $m_A(x)$ us called the ***membership degree*** of *x* in *A*. The greater is $m_A(x)$, the better *x* satisfies the characteristic property of *A*. The definition of the membership function is not unique depending on data, which are usually obtained by statistical or empirical observations. However, a necessary condition for a FS to model in a reliable way the corresponding real situation is that its membership function's definition is compatible to the common sense.

Note that, for reasons of simplicity, many authors identify a FS with its membership function. When U is a finite set, a FS A on U is frequently represented by a symbolic sum in the form $A = \sum_{x \in U} m_A(x)/x$, in which the elements of U having zero membership degrees are omitted. In the same way, if U is a countable set, then A can be written in the form of a symbolic power series, while if U has the power of the continuous, then it can be written as a symbolic integral in the form $\int_{x \in U} m_A(x)dx$.

Example: Let U be the set of non negative integers x such that $x \leq 120$. Then, assuming that the elements of U represent human ages, one can define the fuzzy set A *"The young inhabitants of Greece"* on U by taking its membership function to be

$$m_A(x) = \begin{cases} [1+(0.04x)^2]^{-1}, & x \leq 65 \\ 0, & x > 65 \end{cases}.$$

A crisp subset A of U can be considered as a fuzzy set in U with $m_A(x) = 1$, if $x \in A$ and $m_A(x) = 0$, if $x \notin A$. In this way, most properties and operations of crisp sets can be extended to corresponding properties and operations of fuzzy sets.

For instance, if A, B are fuzzy sets on U, we say that B is a ***subset*** of A and we write $A \subseteq B$, if, and only if, $m_A(x) \leq m_B(x)$, for all x in U. In the special case where A, B are crisp sets, it becomes evident that the above definition coincides with the classical definition of subsets. In the same way, one can define the ***union*** $A \cup B$ and ***intersection*** $A \cap B$ of two fuzzy sets to be the fuzzy sets with membership functions $m_{A \cup B}(x) = max\{m_A(x), m_B(x)\}$ and $m_{A \cap B}(x) = min\{m_A(x), m_B(x)\}$ respectively, etc

For general facts on fuzzy sets we refer to the book of Klir & Folger (1988).

FUZZY LOGIC: AN OVERVIEW

There used to be a tradition in science and engineering of turning to probability theory when one is faced with a problem in which uncertainty plays a significant role. This was necessary when there were no alternative tools for dealing with the uncertainty. However, today this is no longer the rule due to the development of ***Fuzzy logic (FL)*** providing a rich and meaningful addition to the classical logic. Unlike to classical logic, which has only two states true or false, FL deals with truth values which range continuously from 0 to 1. Thus, something could be *half true* (0.5) or *very likely true* (0.9} or *probably not true* (0.1), etc.

The history shows some traces of foundational ideas of FL in the philosophical thoughts put forth by ***Buddha*** around 500 BC. His philosophy was based on the thought that almost everything contains some of its opposite, or in other words that things can be true and false at the same time. However, it was the ancient Greek philosopher ***Plato*** (427-347 BC) who laid the foundation for what would become the FL, by claiming that there exists a third region between true and false, where these opposites "tumbled about". Many centuries later, Hegel, Marx, Engels and other philosophers echoed Plato's sentiments, although ***Lukasiewicz*** was the first who proposed a systematic alternative to the Aristotle's bi-valued logic. In the early 1900's he described the 3-valued logic by adding the term "possible" between Aristotle's 'true" and "false". Later he explored 4 and 5-valued logics and finally he claimed that in principle there was nothing to prevent the derivation of an infinite-valued logic (Lejewski, 1967).

Nevertheless, it was not until recently that the notion of an infinite-valued logic took hold in terms of ***Zadeh's*** (1965) FS theory and in extension of FL. New operations for the calculus of logic were proposed and FL showed to be in principle at least a generalization of the classical logic (Zadeh, 1965, 1968, etc.).

FL allows one to express knowledge in a rule format that is close to a natural language expression and therefore it opens the door to construction of mathematical solutions of problems which are imprecisely defined. Consequently, FL has a much higher PS capability than probability theory, which cannot obtain solutions for such kind of problems.

One must emphasize that, although probabilities and fuzzy membership degrees are both taking values in the interval [0, 1], they are ***distinct*** and ***different*** to each other notions. For example, the expression "*the probability for Mary to be tall is 85%*", means that, although, under the

classical logic's ***principle of the excluded middle*** , Mary is either tall or low in stature, the possibility to be tall, when she is randomly picked up from a particular set of girls, is high. On the contrary the FL expression "*the membership degree of Mary to be tall is 0.85*", simply means that Mary, according to the existing standards for the girls of her age, could be considered as rather tall. Another characteristic difference between probabilities and fuzzy membership degrees is that, while the sum of probabilities of all singleton subsets (events) of *U* equals 1, this is not necessary to happen for the membership degrees.

Another advantage of FL is that, apart of the provision of quantitative information, it gives also the opportunity for a qualitative study of a system's behaviour by examining the behaviour of all possible profiles of the system's units that are involved in a certain activity within the system.

The applications which may be generated from or adapted to FL are wide-ranging covering almost all sectors of human activity. A great variety of systems may be modelled, simulated and even replicated with the help of FL, many of which are connected to the human reasoning its self. This has given in past to many researchers, including the present author and his collaborators, the impulsion to utilize principles of FL for describing in a more effective way the human behaviour in situations characterized by a degree of vagueness and / or uncertainty.

DECISION- MAKING IN FUZZY ENVIRONMENT

In many situations of our day to day life a DM problem is expressed in an ambiguous way involving a degree of uncertainty. In such cases, while the classical statistical decision theory (see Section 3.1) cannot offer an effective help for the DM process, FL due to its nature of including multiple values, offers a rich field of resources. The following two examples illustrate a standard method of DM under fuzzy conditions:

Example 1: A company wants to employ as a sales manager the candidate having the best qualifications, provided that his/her request for salary is not very high and that his/her residence is in a close driving distance from the company's place. There are four candidates for the position, say A, B, C and D with annual salary demands 29050, 25000, 14050, and 6250 euro respectively. Who is the best candidate for the company?

DM process: In the above DM problem we have the ***fuzzy goal (G)*** of employing the best candidate under the ***fuzzy constraints*** that his/her request for salary must not be very high (**C_1**) and that his/her residence is in a close driving distance from the companies place (**C_2**). The steps of the DM process in such vague situations are the following:

Step 1: Choice of the universal set of the discourse

In our case the universal set is the set U = {A, B, C, D} of the four candidates.

Step 2: Fuzzification of the problem's data

In this step the fuzzy goal and the fuzzy constraints of the problem are expressed as fuzzy sets in U. For this, one must define properly the corresponding membership functions. For example, the membership function $m_{C_1}: U \to [0,1]$ for the fuzzy constraint C_1 can be defined by: $m_{C_1}(x) = 1$ for $s(x) < 6000$, $m_{C_1}(x) = 1 - 2 * 10^{-5} * s(x)$ for $6000 \leq s(x) \leq 30000$ and $m_{C_1}(x) = 0$ for $s(x) > 30000$, where $s(x)$ denotes the salary of the candidate x, for all x in U. Then $m_{C_1}(A) = 1 - 2 * 0.2905 = 0.419$. Similarly one finds that $m_{C_1}(B) = 0.5$, $m_{C_1}(C) = 0.719$ and $m_{C_1}(D) = 0.875$. Then the constraint C_1 can be written as a fuzzy set in U in the form of the symbolic sum C_1 = 0.419/A + 0.5/B + 0.719/C + 0.875/D.

Assume further that **G** = 0.9/A + 0.6/B + 0.8/C + 0.6/D and **C_2** = 0.1/A + 0.9/B + 0.7/C + 1/D.

Step 3: Evaluation of the fuzzy data

According to the ***Bellman-Zadeh's*** (1970) criterion for DM in a fuzzy environment the ***fuzzy decision F*** *is the intersection of the fuzzy sets* ***G***, ***C_1*** *and* ***C_2*** *and the solution of the problem corresponds to the element x of U having the highest membership degree in* ***F***.

The membership function of the intersection $G \cap C_1 \cap C_2$ is defined by

$m_{G \cap C_1 \cap C_2}(x) = m_F(x) = \min\{m_G(x), m_{C_1}(x), m_{C_2}(x)\}$ for all x in U. Therefore it is easy to check that $\boldsymbol{F}$ = 0.1/A + 0.5/B + 0.7/C + 0.6/D.

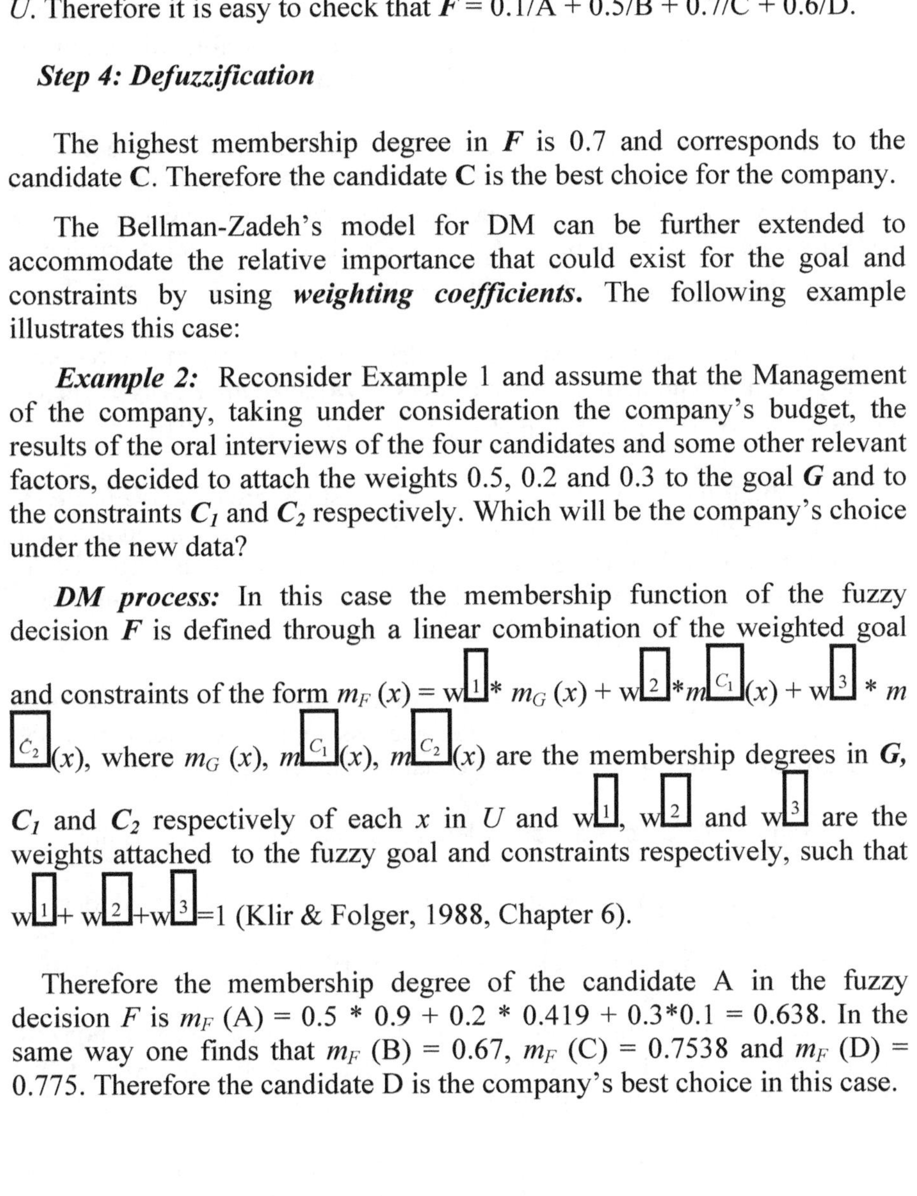

Step 4: Defuzzification

The highest membership degree in $\boldsymbol{F}$ is 0.7 and corresponds to the candidate **C**. Therefore the candidate **C** is the best choice for the company.

The Bellman-Zadeh's model for DM can be further extended to accommodate the relative importance that could exist for the goal and constraints by using ***weighting coefficients.*** The following example illustrates this case:

Example 2: Reconsider Example 1 and assume that the Management of the company, taking under consideration the company's budget, the results of the oral interviews of the four candidates and some other relevant factors, decided to attach the weights 0.5, 0.2 and 0.3 to the goal $\boldsymbol{G}$ and to the constraints $\boldsymbol{C_1}$ and $\boldsymbol{C_2}$ respectively. Which will be the company's choice under the new data?

DM process: In this case the membership function of the fuzzy decision $\boldsymbol{F}$ is defined through a linear combination of the weighted goal and constraints of the form $m_F(x) = w_1 * m_G(x) + w_2 * m_{C_1}(x) + w_3 * m_{C_2}(x)$, where $m_G(x)$, $m_{C_1}(x)$, $m_{C_2}(x)$ are the membership degrees in $\boldsymbol{G}$, $\boldsymbol{C_1}$ and $\boldsymbol{C_2}$ respectively of each x in U and w_1, w_2 and w_3 are the weights attached to the fuzzy goal and constraints respectively, such that $w_1 + w_2 + w_3 = 1$ (Klir & Folger, 1988, Chapter 6).

Therefore the membership degree of the candidate A in the fuzzy decision F is m_F (A) = 0.5 * 0.9 + 0.2 * 0.419 + 0.3*0.1 = 0.638. In the same way one finds that m_F (B) = 0.67, m_F (C) = 0.7538 and m_F (D) = 0.775. Therefore the candidate D is the company's best choice in this case.

EVALUATION OF FUZZY DATA

In Voskoglou (2012), adapting standard rules and principles of FL (Klir & Folger, 1988, Chapter 6), we have studied the behavior of a system's n objects (consisting units), $n \geq 2$, during an activity performed within the system, which involves vagueness and/or uncertainty.

For this, denote by S_i, $i = 1, 2, 3$ the main steps of the activity and by $a, b, c, d,$ and e the ***fuzzy linguistic labels (characterizations)*** of very low, low, intermediate, high and excellent performance respectively of a system's object in each step. Set $U = \{a, b, c, d, e\}$ and attach to each step S_i a fuzzy subset, A_i of U of the form $A_i = \{(x, m_{Ai}(x)): x \in U\}$, $i = 1, 2, 3$.

In order to represent all possible ***profiles (overall states)*** of the system's n objects during the activity we introduce a ***fuzzy relation***, say R, in U^3, which is a FS on U^3 of the form: $R = \{(s, m_R(s)): s=(x, y, z) \in U^3\}$.

Assume that the steps of the activity under study are *depended to each other*. This means that the degree of a system's object success in a certain step depends upon the degree of its success in the previous steps. Under this hypothesis and in order to determine properly the membership function $m_R(s)$ we introduce the following definition:

Definition: A profile $s=(x, y, z)$, with x, y, z in U, is said to be ***well ordered*** if x corresponds to a degree of success equal or greater than y and y corresponds to a degree of success equal or greater than z.

For example, (c, c, a) is a well ordered profile, while (b, a, c) is not.

We define now the membership degree of a profile s to be equal to

$m_R(s) = m_{A_1}(x)\ m_{A_2}(y)\ m_{A_3}(z)$, if s is well ordered, and zero otherwise.

In fact, if for example the profile (b, a, c) possessed a nonzero membership degree, how it could be possible for a system's object that failed during the middle step, to perform satisfactorily at the next step of the activity?

Next, for reasons of brevity, we shall denote the membership degree $m_R(s)$ by m_s. Then the ***fuzzy probability*** p_s of the profile s is defined by $p_s =$

$\frac{m_s}{\sum_{s \in U^3} m_s}$. One can also define the ***possibility*** r_s of s by $r_s = \frac{m_s}{\max\{m_s\}}$, where $max\ \{m_s\}$ denotes the maximal value of m_s, for all s in U^3. Therefore, the possibility of s expresses the "*relative membership degree*" of s with respect to $max\ \{m_s\}$. Shackle (1961) argued that human cognition could be studied better by using possibilities instead of probabilities and many researchers agree nowadays with his view.

Further, in order to study the *combined results* of behaviour of k different groups of the system's objects, $k \geq 2$, during the same activity within the system, we introduce the ***fuzzy variables*** $A_1(t)$, $A_2(t)$ and $A_3(t)$, $t = 1, 2,..,k$. The values of these variables represent the fuzzy subsets of U corresponding to each step of the activity for each of the k groups; e.g. $A_1(2)$ represents the fuzzy subset of U for the second group ($t=2$)corresponding to the first step S_1.

For measuring the degree of evidence of the combined results of the k groups, it becomes necessary to define the probability $p(s)$ and the possibility $r(s)$ of each profile s with respect to the membership degrees of s for all groups. For this, we introduce the ***pseudo-frequencies*** $f(s) = \sum_{t=1}^{k} m_s(t)$ and we define the ***combined probability*** and ***possibility*** of a profile s by $p(s) = \frac{f(s)}{\sum_{s \in U^3} f(s)}$ and $r(s) = \frac{f(s)}{\max\{f(s)\}}$ respectively, where $max\{f(s)\}$ denotes the maximal pseudo-frequency.

Obviously the same method can be applied for studying the combined results of behaviour of a single group of a system's objects during k different activities within the system.

AN APPLICATION TO MARKET'S RESEARCH

The following application was developed in Voskoglou (2003).

An enterprise wants to evaluate the data of a market's research about the consumer acceptance of its products, the level of which is characterized by the ***fuzzy linguistic labels*** of a = negligible, b = low, c = intermediate, d = high and e = very high. The research has been made separately for men and women consumers and for three different categories of age, namely 18-30, 31-50 and over 50 years old.

Set U = {a, b, c, d, e}.The consumer acceptance for each of the above categories of age is represented by a fuzzy set $A_i = \{(x, m_{Ai}(x)), x \in U\}$, i = 1, 2, 3 of U. In order to cover separately men and women, we introduce a parameter t taking the values 1 and 2 respectively. Thus we have the fuzzy variables $A_i(t)$ of U, in which the membership degree of the elements of U depends on the value of t .The fuzzy data of the market's research are depicted in Table 1 .

Table 1: Fuzzy data of the market's research

	$A_1(t)$	$A_2(t)$	$A_3(t)$
t = 1	(0,486 / c) + (0,228 / d) + +(0,286 / e)	(0,171/ a) + (0,171/ b) + +(0,4 / c) + (0,257 / d)	(0,343 / a) + +(0,286 / b) + +(0,371/ c)
t = 2	(0,2 / b) + (0,5 / c) + +(0,3 / d)	(0,2 / a) + (0,267 / b) + +(0,533 / c)	(0,4 / a) + +(0,3 / b) + +(0,3 / c)

Consider the fuzzy relation R = $\{(s, m_R(s)): s = (x, y, z) \in U^3\}$ with membership function $m_R(s) = m_{A_1}(x) m_{A_2}(y) m_{A_3}(z) = m_s$, for each s in U^3. In this case the steps of the activity under study are independent to each other, therefore there is no need to consider here well ordered profiles, as we did in the previous Section. The probabilities and possibilities of the profiles having non zero pseudo- frequencies are depicted in Table 2.

Adding the entries of the fourth and fifth rows of Table 2 one finds that $\sum m_s(1)$=1.049 and $\sum m_s(2)$= 0.939. Therefore $\sum f(s)$= 2.088. It turns out that the profile s = (*c, c, a*) has the greatest pseudo-frequency equal to 0.174 In fact, $m_s(1) = m_{A_1}(c) m_{A_2}(c) m_{A_3}(a)$ =0.486*0.4*0.343 = 0.067, $m_s(2)$ =0.5*0.533*0.4 = 0.107 and *f(s)* = 0.067 + 0.107 = 0.174. Therefore (*c, c, a*) has also the greatest probability of appearance, which is equal to $p(s) = \boxed{\dfrac{0.174}{\sum_{s\in U^3} f(s)}} = \boxed{\dfrac{0.174}{2.088}} = 0.083$ or 8.3%, while its possibility is equal to 1.

Table 2: Profiles with non zero pseudo- frequencies

A_1	A_2	A_3	$m_s(1)$	$m_s(2)$	f(s)	p(s)	r(s)
b	*b*	*b*	0	0.016	0.016	0.008	0.092
b	*a*	*b*	0	0.012	0.012	0.006	0.069
b	*c*	*b*	0	0.032	0.032	0.015	0.184
b	*b*	*a*	0	0.021	0.021	0.010	0.121
b	*b*	*c*	0	0.016	0.016	0.008	0.092
b	*a*	*a*	0	0.016	0.016	0.008	0.092
b	*a*	*c*	0	0.012	0.012	0.006	0.069
b	*c*	*a*	0	0.042	0.042	0.020	0.241
b	*c*	*c*	0	0.032	0.032	0.015	0.184
c	*c*	*c*	0.072	0.080	0.152	0.073	0.874
c	*a*	*c*	0.082	0.030	0.112	0.054	0.644
c	*b*	*c*	0.031	0.040	0.071	0.034	0.408
c	*d*	*c*	0.046	0	0.046	0.022	0.264
c	***c***	***a***	**0.067**	**0.107**	**0.174**	**0.083**	**1**
c	*c*	*b*	0.056	0.008	0.064	0.031	0.368
c	*a*	*a*	0.028	0.040	0.068	0.033	0.391
c	*a*	*b*	0.024	0.030	0.054	0.026	0.310
c	*b*	*a*	0.028	0.053	0.081	0.038	0.466
c	*b*	*b*	0.024	0.040	0.064	0.031	0.368
c	*d*	*a*	0.043	0	0.043	0.021	0.247
c	*d*	*b*	0.036	0	0.036	0.017	0.207
d	*d*	*a*	0.020	0	0.020	0.010	0.115
d	*d*	*b*	0.017	0	0.017	0.008	0.098
d	*d*	*c*	0.022	0	0.022	0.011	0.126
d	*a*	*a*	0.013	0.024	0.037	0.018	0.213
d	*a*	*b*	0.011	0.018	0.029	0.014	0.167
d	*a*	*c*	0.015	0.018	0.033	0.016	0.190
d	*b*	*a*	0.013	0.032	0.045	0.022	0.259
d	*b*	*b*	0.011	0.024	0.035	0.017	0.201
d	*b*	*c*	0.014	0.024	0.038	0.018	0.218
d	*c*	*a*	0.031	0.064	0.095	0,045	0.546
d	*c*	*b*	0.026	0.048	0.074	0.035	0.425
d	*c*	*c*	0.034	0.048	0.082	0.039	0.471
e	*a*	*a*	0.017	0	0.017	0.008	0.098
e	*a*	*b*	0.014	0	0.014	0.007	0.080
e	*a*	*c*	0,018	0	0.018	0.009	0.103

e	*b*	*a*	0.017	0	0.017	0.008	0.098
e	*b*	*b*	0.014	0	0.014	0.007	0.080
e	*b*	*c*	0.018	0	0.018	0.009	0.103
e	*c*	*a*	0.039	0	0.039	0.019	0.224
e	*c*	*b*	0.033	0	0.033	0.016	0.190
e	*c*	*c*	0.042	0	0.042	0.020	0.241
e	*d*	*a*	0.025	0	0.025	0.012	0.144
e	*d*	*b*	0.021	0	0.021	0.010	0.121
e	*d*	*c*	0.027	0	0.027	0.013	0.155

EVALUATION OF STUDENT LEARNING SKILLS

As we have seen in Chapter 3, the MC models developed for the mathematical formulation of several student cognitive actions (PS, MM, Learning, Analogical Reasoning, etc), due to the fact that Probability Theory has not the capability to make forecasts about the "acting in the moment" human behavior, are self-restricted to describe the way in which a student is expected to act (*assumed behavior*) and not how he/she really acts in practice. On the contrary, an advantage of FL is that it has not such limitations.

Here we are going to emphasize the importance of the above FL's advantage by applying the model for evaluating fuzzy data developed earlier in this Chapter to the *process of learning a subject matter in the classroom* (Voskoglou, 1999). For this, let us consider a group of *n* students, $n \geq 2$, during the learning process in the classroom. We denote by S_i, i = 1, 2, 3 , the steps of *interpretation*, *generalization* and *categorization* respectively of the learning process and by n_{ia}, n_{ib}, n_{ic}, n_{id} and n_{ie} the number of students that demonstrated negligible, low, intermediate and excellent performance respectively at the step S_i . We define the membership function $m_{Ai}(x)$ in terms of the frequencies, i.e. by $m_{Ai}(x) = \frac{n_{ix}}{n}$, for each *x* in *U*. Therefore the steps S_i of the learning process are represented as fuzzy sets in *U* in the form:

$$A_i = \{(x, \frac{n_{ix}}{n}) : x \in U\}, \ i = 1, 2, 3.$$

The corresponding experiment took place at the Graduate Technological Educational Institute of Western Greece, at the city

Messolonghi, when I was teaching the definite integral to a group of 35 students of the School of Management. In my short introduction I presented the concept of the definite integral through the need of calculating an area under a curve, but I stated the fundamental theorem of the integral calculus, connecting the indefinite with the definite integral of a continuous in a closed interval function, without proof. Then I left my students to work alone and I was inspecting their efforts and reactions, giving them from time to time suitable hints, or instructions. My intension was to help them to understand the basic methods of calculating a definite integral in terms to the already known methods for the indefinite integral (step S_1 of interpretation). I observed that 17, 8 and 10 students achieved intermediate, high and excellent understanding respectively of the new subject matter. Therefore,
$n_{1a} = n_{1b} = 0$, $n_{1c} = 17$, $n_{1d} = 8$ and $n_{1e} = 10$, which means that S_1 is represented as a fuzzy set on U as $A_1 = \{(a,0),(b,0),(c,\frac{17}{35}),(d,\frac{8}{35}),(e,\frac{10}{35})\}$.

In the next step I gave to students for solution a number of exercises and simple problems involving calculations of improper integrals as limits of definite integrals, and of the area under a curve, or among curves. My target was to help them to generalize the new information to a variety of situations (step S_2). In this case I found that $A_2 = \{(a,\frac{6}{35}), (b,\frac{6}{35}), (c,\frac{16}{35}), (d,\frac{7}{35}), (e,0)\}$.

At the final step I gave to students for solution a number of composite problems involving applications to economics, such as calculation of the present value in cash flows, of the consumer's and producer's surplus resulting from the change of prices of a given good, of probability density functions, etc. (Dowling, 1980, chapter 17). My intension was to help students to relate the new information to their existing knowledge structures (step S_3). In this case I found that
$A_3 = \{(a,\frac{12}{35}), (b,\frac{10}{35}), (c,\frac{13}{35}), (d,0), (e,0)\}$.

Observing the *Ai's* one concludes, as it was logically expected, that the higher is the step of the learning process, the lower the student performance.

It is straightforward now to calculate the membership degrees of all possible student profiles during the learning process, which are presented in the column of $m_s(1)$ in Table 1. For example, for $s = (c, b, a)$ one finds that

$$m_s = m_{A_1}(c).\ m_{A_2}(b).\ m_{A_3}(a) = \frac{17}{35}\frac{6}{35}\frac{12}{35} = \frac{1224}{42875} \approx 0.029, \text{ etc.}$$

Shackle (1961) argued that human cognition can be studied more efficiently by possibilities rather, than by probabilities and many researchers nowadays adopt this argument. Here, for calculating the possibilities of student profiles observe first that profile (c, c, c) has the maximal membership degree, equal to 0.082. Consequently, the possibility of each s in U^3 is calculated by the formula $r_s = \frac{m_s}{0.082}$. For example, the possibility of (c, b, a) is equal to $\frac{0{,}029}{0.082} \approx 0.353$, the possibility of (c, c, c) is 1, etc. (see the column of $r_s(1)$ in Table 2).

A few days later I delivered the same lecture to a group of 30 students of another Department of the School of Management. Working in the same way I found that

$$A_1 = \{(a, 0), (b, \tfrac{6}{30}), (c, \tfrac{15}{30}), (d, \tfrac{9}{30}), (e, 0)\},$$

$$A_2 = \{(a, \tfrac{6}{30}), (b, \tfrac{8}{30}), (c, \tfrac{16}{30}), (d, 0), (e, 0)\} \text{ and}$$

$$A_3 = \{(a, \tfrac{12}{30}), (b, \tfrac{9}{30}), (c, \tfrac{9}{30}), (d, 0), (e, 0)\}.$$

Then, I calculated the student profiles, which are presented in the column of $m_s(2)$ in Table 1. It turns out that (c, c, a) is the profile with the maximal membership degree 0.107 and therefore the possibility of each s is calculated by

$r_s = \frac{m_s}{0{,}107}$. The possibilities of all profiles are presented in the column of $r_s(2)$ in Table 3.

Table 3: Student profiles with non zero pseudo-frequencies

A_1	A_2	A_3	$m_s(1)$	$r_s(1)$	$m_s(2)$	$r_s(2)$	$f(s)$	$r(s)$
B	*b*	*b*	0	0	0.016	0.150	0.016	0.087
B	*b*	*a*	0	0	0.021	0.196	0.021	0.115

B	*A*	*a*	0	0	0.016	0.150	0.016	0.087
C	*C*	*c*	0.082	1	0.080	0.748	0.162	0.885
C	*C*	*a*	0.076	0.927	0.107	1	0.183	1
C	*C*	*b*	0.063	0.768	0.008	0.075	0.071	0.388
C	*A*	*a*	0028	0.341	0.040	0.374	0.068	0.372
C	*B*	*a*	0.028	0.341	0.053	0.495	0.081	0.443
C	*B*	*b*	0.024	0.293	0.040	0.374	0.064	0.350
D	*D*	*a*	0.016	0.495	0	0	0.016	0.087
D	*D*	*b*	0.013	0.159	0	0	0.013	0.074
D	*D*	*c*	0.021	0.256	0	0	0.021	0.115
D	*A*	*a*	0.013	0.159	0.024	0.224	0.037	0.202
D	*B*	*a*	0.013	0.159	0.032	0.299	0.045	0.246
D	*B*	*b*	0.011	0.134	0.024	0.224	0.035	0.191
D	*C*	*a*	0.031	0.378	0.064	0.598	0.095	0.519
D	*C*	*b*	0.026	0.317	0.048	0.449	0.074	0.404
D	*C*	*c*	0.034	0.415	0.048	0.449	0.082	0.448
e	*A*	*a*	0.017	0.207	0	0	0.017	0.093
e	*B*	*b*	0.014	0.171	0	0	0.014	0.077
e	*c*	*a*	0.039	0.476	0	0	0.039	0.213
E	*C*	*b*	0.033	0.402	0	0	0.033	0.180
E	*C*	*c*	0.042	0.512	0	0	0.042	0.230
E	*D*	*a*	0.025	0.305	0	0	0.025	0.137
E	*D*	*b*	0.021	0.256	0	0	0.021	0.115
E	*D*	*c*	0.027	0.329	0	0	0.027	0.148

The outcomes in Table 3 have been calculated with accuracy up to the third decimal point.

Next, in order to study the combined results of the behaviour of the two groups, we introduce the fuzzy variables $A_i(t)$, with $i = 1, 2, 3$ and $t = 1, 2$. The pseudo-frequencies of the student profiles, which are calculated by the formula
$f(s) = m_s(1) + m_s(2)$, are also presented in the corresponding column of Table 4. It turns out that *(c, c, a)* is the profile with the highest pseudo-frequency 0.183 and therefore the combined possibilities of the student profiles are calculated by the formula $r(s) = \boxed{\frac{f(s)}{0,183}}$. The combined possibilities of all profiles with non zero pseudo-frequencies are shown in the last column of Table 3.

In concluding, from the classroom experiment described above it becomes evident that the use of the fuzzy model developed in Section 4.4, for the representation of the process of learning, apart from quantitative information (membership degrees and possibilities), gives also a qualitative view of the student behaviour in the classroom by studying all possible student profiles. In this way the instructor gets, in terms of the fuzzy linguistic characterizations used, a comprehensive idea of the degree of acquisition of each step of the learning process by students. Therefore, it becomes much easier for him/her to readapt properly the process, the rate and possibly the methods of his/her lectures in order to succeed. better results in future.

Another advantage of the fuzzy model is that it gives the possibility of studying the combined results of two or more student groups during the learning process of the same subject matter, or alternatively of the same student group during the learning process of two, or more, different subjects. This could be very useful in cases where one wishes to study the overall performance of several school classes, etc.

Finally, it becomes evident that, apart from mathematics, the above model could be used for studying the learning process of any other topic.

.

OTHER APPLICATIONS

The general character of the model for evaluating fuzzy data developed in this Chapter enabled the present author to apply it for the description of other student cognitive actions as well, such as PS, MM, CBR, etc. (see the book of Voskoglou, 2011 and the relevant references contained in it). Note that a similar model has been applied by Perdikaris (2002) for assessing student Geometric skills in terms of the van Hiele levels of geometric reasoning.

CONCLUSION

In this Chapter, apart from the basics from Fuzzy Sets and Logic needed for the rest of the book, the process of Decision-Making in a fuzzy environment is studied and a general model is developed for evaluating a system's fuzzy data with applications to market's research and to the process of learning a subject matter in the classroom.

REFERENCES

Bellman, R. E. & Zadeh, L. A. (1970), Decision making in fuzzy environment, *Management Science*, 17, 141-164.

Klir, G.J. & Folger, T.A. (1988), *Fuzzy Sets, Uncertainty and Information*, Prentice-Hall, London.

Lejewski, C. (1967), Jan Lukasiewicz, *Encyclopedia of Philosophy,* Vol. 5, 414-417,
MacMillan, New York.

Perdikaris, S. (2002), Measuring the student group capacity for obtaining geometric information in the van Hiele development thought process: A fuzzy approach, *Fuzzy Sets and Mathematics*, 16 (3), 81-86.

Shackle, G. L. S. (1961), *Decision, Order and Time in Human Affairs*, Cambridge University Press, Cambridge, 1961

Voskoglou, M. Gr. (1999), The process of learning mathematics: A fuzzy set approach, *Heuristic and Didactics of Exact Sciences*, 10, 9-13.

Voskoglou, M. Gr. (2003), Applications of Fuzzy Sets to Problems of Commercial Enterprises, *Proceedings of 1st International Conference on Quantitative Methods in Industry and Commerce,* 654-659, Graduate T. E. I. of Athens, Greece.

Voskoglou, M. Gr. (2011), *Stochastic and fuzzy models in Mathematics Education, Artificial Intelligence and Management*, Lambert Academic Publishing, Saarbrucken, Germany.

Voskoglou, M. Gr. (2012), A study on fuzzy systems, *American Journal of Computational and Applied Mathematics*, 2(5), 232-240.

Zadeh, L. A. (1965), Fuzzy Sets, *Information and Control*, 8, 338-353.

Zadeh, L. A. (1968), Fuzzy algorithms, *Information and Control,* 12, 94-102.

CHAPTER 5
Measuring Fuzzy System Uncertainty

ABSTRACT

A basic principle of Information Theory states that a system's uncertainty can be viewed as its capacity to obtain information. In this Chapter two types of a fuzzy system's uncertainty are discussed: The probabilistic one, which is an adaptation of the classical Shannon's entropy for functioning in a fuzzy environment and the total possibilistic uncertainty, which according to many researchers is more suitable for the study of human cognitive action. Examples are also presented to illustrate our results and in particular for managing the uncertainty in the van Hiele Levels of Geometric Reasoning.

FUZZY SYSTEMS

In Chapter 1 we have defined a ***system*** as a set of interacting components forming an integrated whole. More explicitly, the word system (from Latin *systema,* in turn of the Greek *σύστημα*) has a long history which can be traced back to Plato's "Phillebus", to Aristotle's "Politics" and to Euclid's "Elements". It had meant "total", "crowd" or "union" in even more ancient times, as it derives from the verb *sunistemi*, which means uniting, putting together. Nowadays, in the most general sense, system means a configuration of parts connected and joined together by a web of relationships.

The first to develop the concept of a system in the natural sciences was the French physicist Carnot in 1824, who studied thermodynamics. In 1850 the German physicist Clausius generalized Carnot's picture to include the concept of the surroundings and began to use the term "working body"

when referring to the system. One of the pioneers of the general systems' theory was the biologist Bertalanffy, while significant development to the concept of the system was done by Wienner and Ashby, who pioneered the use of mathematics to study systems. Contemporary ideas from system theory have grown with diversified areas, exemplified by the works of Banathy (1996), Hammond (2003), Odum (1994) and others.

Currently, applications of the system concept include information and computer science, engineering and physics, social and cognitive sciences, management and economics, strategic thinking, fuzziness and uncertainty, etc. ***System Theory*** thus serves as a bridge for interdisciplinary dialogue between autonomous areas of study, as well as within the area of systems' science itself.

Most systems share common characteristics including structure, behavior, interconnectivity (the various parts of a system have functional and structural relations to each other), sets of functions, etc. We scope a system by defining its boundary; this means choosing which entities are inside the system and which are outside, part of the environment. We then construct ***models*** of the system in order to understand it and to predict or impact its future behavior.

Situations are often appearing in a system's operation characterized by a degree of vagueness. Such a system is usually referred as a ***Fuzzy System.*** For example, during the PS process students frequently utilize concepts or processes which are intuitively graded by them in a fuzzy way. From the teacher's point of view also, there usually exists vagueness about the degree of student understanding of each of the steps of the new didactic object's presentation, etc.

The use of rules and principles of FL is normally required for the construction of a fuzzy system's mathematical model. In Section 4.3 we have already seen an example of ***MM under fuzzy conditions*** for the DM process. This process involves in general the following steps (Klir & Folger, 1988):

- ***Choice of the universal set*** of the discourse.

- ***Fuzzification*** of the problem's data by defining proper membership functions.

- ***Evaluation of the fuzzy data*** by applying rules and principles of FL to obtain a unique FS, which determines the required solution.
- ***Defuzzification*** of the final outcomes in order to apply the solution found to the original, real world problem.

PROBABILISTIC UNCERTAINTY

Uncertainty is the shortage of precise knowledge and of complete information on data, which describe together the state of the corresponding system. One of the key problems of artificial intelligence is the modelling of uncertainty for solving real life problems and several models have been proposed for this purpose.

A basic principle of the classical ***Information Theory*** states that a system's uncertainty can be viewed as its capacity for obtaining information. The more is the reduction of the uncertainty the greater the amount of the new information obtained. Therefore, the measurement of a system's uncertainty could be used for assessing its ***performance***, i.e. its effectiveness to obtain the required amount of information by an action performed within it.

Note that, the assessment of a human-designed system's performance is one of the most important parts of its study helping the system's designer to make all the necessary modifications/improvements to the system's structure in order to increase its effectiveness.

The traditional way of measuring a system's uncertainty is by the Shannon's (1948) formula, obtained in terms of probability theory and usually termed as the ***Shannon's entropy*** [1]. For use in a fuzzy environment the Shannon's entropy has been adapted to the form $H = -\frac{1}{\ln n}\sum_{s=1}^{n} m_s \ln m_s$ (1) (Klir, 1995, p. 20), , where n denotes the total number of the system's objects involved in the corresponding action and m_s is the membership degree of the profile s. The sum is divided by the natural logarithm of n in order to be normalized. Thus H takes values in the interval [0, 1].

Example: Reconsider the classroom experiment for the process of learning a subject matter in the classroom.presented in Chapter 4. Then replacing the values of m_s given in the columns of $m_s(1)$ and of $m_s(2)$ of

Table 3 to equation (1) and making the required calculations one finds the values $H\approx 0.472$ for the first and $H\approx 0.419$ for the second student group respectively.

POSSIBILISTIC UNCERTAINTY

Within the domain of possibility theory uncertainty consists of ***strife (or discord)***, which expresses conflicts among the various sets of alternatives, and ***non-specificity (or imprecision)***, which indicates that some alternatives are left unspecified, i.e. it expresses conflicts among the sizes (cardinalities) of the various sets of alternatives (Klir, 1995, p.28).

For a better understanding of the above two types of uncertainty we present the following simple example:

Example 1: Let U be the set of non negative integers $x\leq 120$ representing human ages and let Y = young, A = adult and O = old be fuzzy subsets of U defined by the membership functions m_Y, m_A and m_O respectively, where people are considered as young, adult or old according to their outer appearance . Then, given x in U, a degree of uncertainty usually exists about the values that the membership degrees $m_Y(x)$, $m_A(x)$ and $m_O(x)$ could take, resulting to a conflict among the fuzzy subsets Y, A and O of U. For instance, if x = 18, values like $m_Y(x)$ = 0.8 and $m_A(x)$ = 0.3 are logically acceptable, but they are not the only ones. In fact, the values $m_Y(x)$ = 1 and $m_A(x)$ = 0.5 are also acceptable, etc. The existing conflict becomes even greater if x = 50. In fact, is it reasonable in this case to take $m_Y(x)$ =0? Probably not, because sometimes people being 50 years old look much younger than others aged 40 or even 30 years. On the other hand, people aged 50 can be found also who look older than others aged 70, or even 80 years! The same question arises about the logically acceptable values of $m_O(x)$. All the above are examples of the type of uncertainty that we have termed as strife.

On the other hand, non - specificity is connected to the question: How many x in U could have non zero membership degrees in Y, A and O respectively? In other words, the existing in this case uncertainty creates a conflict among the cardinalities (sizes) of the fuzzy subsets of U. It is recalled that the ***cardinality*** of a fuzzy subset, say *B*, of *U* is defined by the sum $\sum_{x \in U} m_B(x)$ of all membership degrees of the elements of U in *B*. –

Strife is calculated by the function *ST(r)* = $\frac{1}{\log 2}[\sum_{i=2}^{n}(r_i - r_{i+1})\log\frac{i}{\sum_{j=1}^{i} r_j}]$ (2),

defined on the ***ordered possibility distribution*** *r*: $r_1 = 1 \geq r_2 \geq \ldots\ldots \geq r_n \geq r_{n+1}$ of a system's objects (Klir, 1995, p.28).

In the same way (Klir, 1995, p.28) non-specificity is calculated by the function

$$N(r) = \frac{1}{\log 2}[\sum_{i=2}^{n}(r_i - r_{i+1})\log i] \qquad (3).$$

The sum *T(r)* = *ST(r)* + *N(r)* measures the system's ***total possibilistic uncertainty***, i.e. its capacity to obtain new information.

The following example illustrates the calculation of *T(r)* in practice:

Example 2: Reconsider the classroom experiment for the process of learning a subject matter in the classroom presented in Chapter 4. Observing the values of possibilities of the student profiles of the first group given in the column of $r_s(1)$ of Table 3 one finds that the ordered possibility distribution for this group is *r*: $r_1=1$, $r_2 = 0.927$, $r_3 = 0.768$, $r_4 = 0.512$, $r_5 = 0.476$, $r_6 = 0.415$, $r_7 = 0.402$, $r_8 = 0.378$, $r_9 = r_{10} = 0.341$, $r_{11} = 0.329$, $r_{12} = 0.317$, $r_{13} = 0.305$, r_{14} = 0,293, $r_{15} = r_{16} = 0.256$, $r_{17} = 0.207$, $r_{18} = 0.195$, $r_{19} = 0.171$, $r_{20} = r_{21} = r_{22} = 0.159$, $r_{23} = 0.134$, $r_{24} = r_{25} = \ldots\ldots\ldots\ldots\ldots = r_{125} = 0$.

Replacing the above values to formulas (2) and (3) and making the required calculations one finds that $ST(r) \approx 0.565$ and $N(r) \approx 2.405$, which gives that $T(r) \approx 2.97$.

In the same way, working with the values of possibilities given in the column of $r_s(2)$ of Table 3 one finds that $T(r) \approx 2.322$ for the second student group.

.

MANAGING THE UNCERTAINTY IN THE VAN HIELE LEVELS OF GEOMETRIC REASONING.

The van Hiele Levels of Geometric Reasoning

The pedagogical value of the Euclidian geometry is great, mainly because it cultivates the student cognitive skills and connects directly mathematics to the real world. However, students face many difficulties for the learning of the Euclidian geometry, which fluctuate from the understanding of the space to the development of geometric reasoning and the ability of constructing the proofs and solutions of several geometric propositions and problems.

The ***van Hiele (vH)*** theory of geometric reasoning (van Hiele & van Hiele-Geldov, 1958, van Hiele, 1986, Haniger & Vozkunkova, 2014, Hsiu-Lau Ma et al. 2015, etc.) suggests that students can progress through five levels of increasing structural complexity. A higher level contains all knowledge of any lower level and some additional knowledge which is not explicit at the lower levels. Therefore, each level appears as a meta-theory of the previous one (Freudenthal, 1973). The five vH levels include:

- L_1 ***(Visualization):*** Students perceive the geometric figures as entities according to their appearance, without explicit regard to their properties.

- L_2 ***(Analysis):*** Students establish the properties of geometric figures by means of an informal analysis of their component parts and begin to recognize them by their properties.

- L_3 ***(Abstraction):*** Students become able to relate the properties of figures, to distinguish between the necessity and sufficiency of a set of properties in determining a concept and to form abstract definitions.

- L_4 ***(Deduction):*** Students reason formally within the context of a geometric system and they gasp the significance of deduction as means of developing geometric theory.

- L_5 ***(Rigor):*** Students understand the foundations of geometry and can compare geometric systems based on different axioms.

Obviously the level L_5 is very difficult, if not impossible, to appear in secondary classrooms, whereas level L_4 also appears very rarely.

Although van Hiele (1986) claimed that the above levels are discrete – which means that the transition from a level to the next one does not happen gradually but all at once – alternative researches by Burger & Shaughnessy (1986), Fuys et al. (1988), Wilson (1990), Guttierrez et al. (1991) and by Perdikaris (2011) suggest that the vH levels are ***continuous*** characterized by transitions between the adjacent levels. This means that from the teacher's point of view there exists fuzziness about the degree of student acquisition of each vH level. Therefore, principles of FL can be used for the assessment of student geometric reasoning skills.

Measuring the Student Uncertainty in the van Hiele Levels

Gutierrez et al. (1991) presented a paradigm for evaluating the acquisition of the vH levels in three–dimensional Geometry by three different groups, say G_1, G_2 and G_3, consisting of 20, 21 and 9 high school students respectively. Here, we shall use the data of this paradigm, which are depicted in Table 1, to measure the uncertainty of the student groups.

Table 1: Data of the Gutierrez et al. (1991) paradigm

Group	vH level	F	D	C	B	A
G_1	L_1	0	0	0	0	20
G_1	L_2	1	0	3	6	10
G_1	L_3	2	3	6	6	3
G_2	L_1	0	0	1	2	18
G_2	L_2	0	3	4	13	1
G_2	L_3	9	6	5	1	0
G_3	L_1	0	2	4	2	1
G_3	L_2	3	4	2	0	0
G_3	L_3	9	0	0	0	0

The letters A, B, C, D and F in Table 1 correspond to the fuzzy linguist labels (grades) of excellent, very good, good, fair and non satisfactory acquisition respectively of the vH levels of geometric reasoning by students. Therefore, the set of the discourse in this paradigm is the set U = {A, B, C, D, F}.

The data of Table 1 are characterized by a significant degree of fuzziness, because the student performance is not evaluated by numerical scores, but in terms of the fuzzy grades of U. Consequently, the traditional method of calculating the mean value of the student scores cannot be applied here for assessing the student group mean performance. Instead,

we shall calculate the probabilistic and total possibilistic uncertainty of the student groups.

i) Calculation of total possibilistic uncertainty:

Perdikaris (2002) used fuzzy possibilities and the student group total possibilistic uncertainty to compare the intelligence of student groups in the vH level theory. He considered all the profiles of the form (x, y, z) with x, y and z in U representing a student's performance in the vH levels L_1, L_2 and L_3 respectively and he defined the membership degrees of those profiles by the product $\frac{n_x}{n} \cdot \frac{n_y}{n} \cdot \frac{n_z}{n}$ of the corresponding frequencies, where the set of all profiles is represented as a fuzzy subset of U^3.

However, the above definition is problematic, because it assigns non – zero membership degrees to profiles like (A, B, A), (B, A, D), etc. In other words, in those profiles the student's performance in a vH level is assumed to be worse than that in the next level, which is impossible to happen. This problem was resolved by Voskoglou (2009), who developed a similar model for the process of learning, (see Chapter 4) by assigning non-zero membership degrees only to well defined student profiles.

The above method, although it performs a useful for the instructor/researcher quantitative analysis of all student profiles in terms of their possibilities, it is very laborious even in its revised form (Voskoglou, 2009) requiring the calculation of the membership degrees of 5^3 in total student profiles - ordered samples of three taken from the five in total elements of U with replacement - and the corresponding possibilities by the formulas given in the previous Chapter.

Here we develop a much simpler variation of this method (Voskoglou, 2017). For this, we represent the three student groups as fuzzy subsets of U defining their membership functions in terms of the frequencies. Under this manipulation the data of the first row of Table 1, concerning the performance of G_1 in level L_1, imply that the ordered possibility distribution is $r_1=r(A)=1>r_2=r_3=r_4=r_5=0$ and therefore from formulas (2) and (3) one finds immediately that the total possibilistic uncertainty for G_1 in L_1 is $T_1(r)=0$.

Consider now the data of the second row of Table 1 concerning the performance of G_1 in level L_2. In this case one finds the membership degrees $m(F)=\frac{1}{20}$, $m(D)=0$, $m(C)=\frac{3}{20}$, $m(B)=\frac{6}{20}$ and $m(A)=\frac{10}{20}$. Since $m(A)$ is the maximal membership degree, the corresponding ordered possibility distribution is $r_1=r(A)=1>r_2=r(B)=\frac{6}{10}>r_3=r(C)=\frac{3}{10}>r_4=r(F)=\frac{1}{10}>r_5=r(D)=0$. Therefore, since $r_{n+1}=r_5$, i.e. $n=4$, formula (2) gives that

$$ST(r)=\frac{1}{\log 2}[(r_2-r_3)\log\frac{2}{r_1+r_2}+(r_3-r_4)\log\frac{3}{r_1+r_2+r_3}+(r_4-r_5)\log\frac{4}{r_1+r_2+r_3+r_4}]=$$

$$\frac{1}{\log 2}[\frac{3}{10}\log(\frac{20}{16})++\frac{2}{10}\log(\frac{30}{19})+\frac{1}{10}\log(\frac{40}{20})]\approx 0.29.$$

Also, formula (3) gives that $N(r)=\frac{1}{\log 2}[\frac{3}{10}\log 2+\frac{2}{10}\log 3+\frac{1}{10}\log 4]\approx 0.82$. Therefore, the total possibilistic uncertainty for G_1 in L_2 is $T_2(r)\approx 1.11$.

Next, from the third row of Table 1 one finds for G_1 in level L_3 that $m(F)=\frac{2}{20}$, $m(D)=\frac{3}{20}$, $m(C)=m(B)=\frac{6}{20}$ and $m(A)=\frac{3}{20}$. Consequently, we have the ordered possibility distribution $r_1=r_2=1>r_3=r_4=\frac{3}{6}>r_5=\frac{2}{6}$ wherefrom we find that

$$ST(r)=\frac{1}{\log 2}[\frac{3}{6}\log(\frac{2}{2})+\frac{1}{6}\log(\frac{4}{3})]\approx 0.69,\ N(r)=\frac{1}{\log 2}(\frac{3}{6}\log 2+\frac{1}{6}\log 4)\approx 0.83.$$

Therefore, for G_1 in L_3 we have $T_3(r)\approx 1.52$.

It is logical now to accept that the mean value $T(r) = \frac{T_1(r)+T_2(r)+T_3(r)}{3} \approx 0.88$ measures the total possibilistic uncertainty for G_1 in the first three vH levels L_1, L_2 and L_3.

In the same way one finds from the fourth row of Table 1 the ordered possibility distribution $r_1=1>r_2=\frac{2}{18}>r_3=\frac{1}{18}>r_4=r_5=0$, which gives that $ST(r) \approx 0.12$, $N(r) \approx 0.14$ and $T_1(r) \approx 0.26$ for G_2. Further, the fifth row gives $r_1=1>r_2=\frac{4}{13}>r_3=\frac{3}{13}>r_4=\frac{1}{13}>r_5=0$, $ST(r) \approx 0.3$, $N(r) \approx 0.48$ and $T_2(r) \approx 0.78$. Also the sixth row gives $r_1=1>r_2=\frac{6}{9}>r_3=\frac{5}{9}>r_4=\frac{1}{9}>r_5=0$, $ST(r) \approx 0.31$, $N(r) \approx 1.04$ and $T_3(r) \approx 1.35$. Therefore, the total possibilistic uncertainty for G_2 is equal to $T(r) = \frac{0.26+0.78+1.35}{3} \approx 0.8$.

Similarly, from the last three rows of Table 1 one finds that for G_3 we have

$$T(r) = \frac{1.27+1.68+0}{3} \approx 0.98$$

ii) Calculation of the probabilistic uncertainty: We consider again the three student groups as fuzzy subsets of U. Replacing the data of the first row of Table 5 to formula (1) one finds that the probabilistic uncertainty for G_1 in level L_1 is $H_1 = 0$. Also, replacing the membership degrees of the elements of U for G_1 in L_2 calculated in paragraph (i) to formula (1) one finds that

$$H_2 = -\frac{1}{\ln 5}\left(\frac{10}{20}\ln\frac{10}{20}+\frac{6}{20}\ln\frac{6}{20}+\frac{3}{20}\ln\frac{3}{20}+\frac{1}{20}\ln\frac{1}{20}\right) \approx 0.89.$$

Further, replacing the membership degrees for G_1 in L_3 to formula (1) one finds that $H_3 \approx 0.95$. Hence, the probabilistic uncertainty for G_1 for the first three vH levels could be considered to be $H = \frac{H_1+H_2+H_3}{3} \approx 0.61$.

In the same way one calculates the probabilistic uncertainty for the groups G_2 and G_3.

OTHER APPLICATIONS

In earlier author's works the measurement of a fuzzy system's uncertainty has been also applied as an assessment method of the system's performance in a variety of other cases, such as PS, MM, CBR with the help of computers, etc. For example, see the books Voskoglou, 2011 and Voskoglou, 2016: Chapter 5 and the relevant references contained in them.

CONCLUSIONS

The following conclusions can be drawn from the discussion performed in this Chapter:

- A system's uncertainty can be considered as its capacity to obtain information.

- Two methods were discussed for measuring a fuzzy system's uncertainty. The first one calculates the system's ***probabilistic uncertainty*** by applying the classical Shannon's entropy properly adapted for use in a fuzzy environment. The second method calculates the system's ***total possibilistic uncertainty***, which is equal to the sum of ***strife*** and of ***non specificity***.

- The first method calculates the probabilistic uncertainty directly from the membership degrees of the corresponding profiles. On the contrary, the second method requires the calculation of the profile order possibility distribution first, which makes it more laborious. Nevertheless, adopting Shackle's (1961) view that human cognition can be studied better by the possibility rather than by the probability theory it is suggested to use the second method in case of human spiritual activities.

REFERENCES

Banathy, B. (1996), *Designing Social Systems in Changing World,* Plenum, New York.

Burger, W.P. & Shaughnessy, J.M. (1986), Characterization of the vam Hiele Levels of Development in Geometry, *Journal for Research in*

Mathematics Education, 17, 31-48.

Freudenthal, H. (1973), *Mathematics as an Educational Task,* D. Reidel, Dordrecht.

Fuys, D., Geddes, D. & Tischler, R. (1988), *The van Hiele Model of Thinking in Geometry among Adolescents,* Journal for Research in Mathematics Education, Monograph 3, NCTM, Reston, VA, USA.

Gutierrez, A., Jaine, A. & Fortuny, J.K. (1991), An Alternative Paradigm to Evaluate the Acquisition of the van Hiele Levels, *Journal for Research in Mathematics Education,* 22, 237-251.

Hammond, D. (2003), *The Science of Synthesis*, University of Colorado Press, Colorado.

Haviger, J. & Vozkunkova, I. (2014), The van Hiele geometry thinking levels: Gender and school type differences, *Procedia – Social and Behavioral Sciences,* 112, 977-981.

Hsiu-Lan Ma et al. (2015), A study of van Hiele geometric thinking among 1st through 6th Grades, *Eurasia Journal of Mathematics Science ant Technical Education,* 11(5), 1181-1196.

Klir, G.J. & Folger, T.A. (1988), *Fuzzy Sets, Uncertainty and Information*, Prentice-Hall, London.

Klir, G.J. (1995), Principles of Uncertainty: What are they? Why do we mean them?, *Fuzzy Sets and Systems*, 74, 15-31.

Odum, H. (1994), Ecological and General Systems: An introduction to systems ecology, Colorado University Press, Colorado.

Perdikaris, S. (2002), Measuring the student group capacity for obtaining geometric information in the van Hiele development thought process: A fuzzy approach, *Fuzzy Sets and Mathematics*, 16 (3), 81-86.

Perdikaris, S.C. (2011), Using Fuzzy Sets to Determine the Continuity of the van Hiele Levels, *Journal of Mathematical Sciences and Mathematics Education,* 6(3), 81-86.

Shackle, G. L. S. (1961), *Decision, Order and Time in Human Affairs*, Cambridge University Press, Cambridge, 1961

Shannon, C. E. (1948), A mathematical theory of communications, *Bell Systems Technical Journal*, 27, 379-423 and 623-656.

van Hiele, P.M. & van Hiele-Geldov, D. (1958), *Report on Methods of Initiation into Geometry,* edited by H. Freudenthal, J.B. Wolters,

Groningen, The Netherlands, pp. 67-80.

van Hiele, P.M. (1986), *Structure and Insight,* Academic Press, New York.

Voskoglou, M. Gr. (2009), Fuzziness or Probability in the Process of Learning: A General Question Illustrated by Examples from Teaching Mathematics, *The Journal of Fuzzy Mathematics,* International Fuzzy Mathematics Institute (Los Angeles), 17(3), 679-686.

Voskoglou, M. Gr. (2011), *Stochastic and fuzzy models in Mathematics Education, Artificial Intelligence and Management*, Lambert Academic Publishing, Saarbrucken, Germany.

Voskoglou, M. Gr. (2016), *Finite Markov Chain and Fuzzy Models in Management and Education,* GIAN Program, Course No. 16102K03/2015-16, National Institute of Technology, Durgapur, India

Voskoglou, M. Gr. (2017), Managing the Uncertainty in the van Hiele Levels of Geometric Reasoning, *American Journal of Educational Research*, 5(2), 109-113.

Wilson, M. (1990), Measuring a van Hiele Geometric Sequence: A Reanalysis, *Journal for Research in Mathematics Education,* 21, 230-237.

ENDNOTE

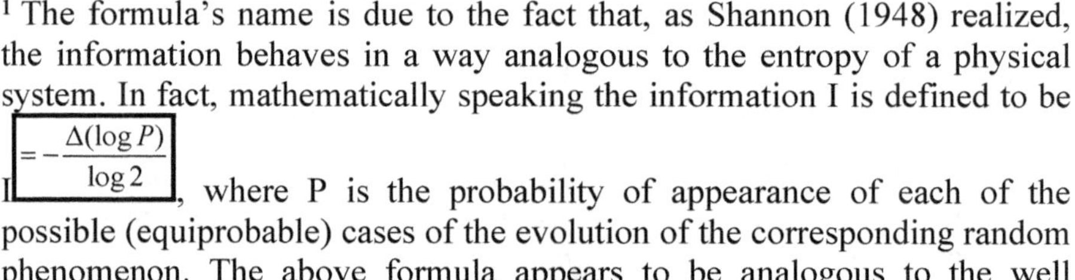

[1] The formula's name is due to the fact that, as Shannon (1948) realized, the information behaves in a way analogous to the entropy of a physical system. In fact, mathematically speaking the information I is defined to be I $= -\frac{\Delta(\log P)}{\log 2}$, where P is the probability of appearance of each of the possible (equiprobable) cases of the evolution of the corresponding random phenomenon. The above formula appears to be analogous to the well known from Physics formula $\Delta S = \frac{\Delta Q}{T}$, where ΔS is the increase of a physical system's entropy caused by an increase of the heat by ΔQ, when the absolute temperature T remains constant. In other words, the information acts as a kind of "negative entropy" (***negentropy***) of the corresponding system.

CHAPTER 6
The Centre of Gravity Technique and its Variations

ABSTRACT

The Centre of Gravity (COG) Technique is the most popular defuzzification method of fuzzy mathematics. In this Chapter the COG technique is properly adapted to create the Rectangular Fuzzy Assessment Model (RFAM). Variations of the RFAM are also studied including the Generalized Rectangular Fuzzy Assessment Model (GRFAM), the Triangular Fuzzy Assessment Model (TFAM) and the Trapezoidal Fuzzy Assessment Model (TpFAM). Nevertheless, as it is finally proved, all these models are equivalent in practice, since they provide the same assessment outcomes when comparing the performance of different groups. The use of the above models is illustrated by applying them for student and Bridge player assessment, while other applications are also mentioned developed in earlier works. The RFAM and its variations are validated by comparing their outcomes with the corresponding outcomes of two traditional assessment methods of the bi-valued logic, the calculation of the mean values and the Grade Point Average (GPA) index

---

TRADITIONAL ASSESSMENT METHODS

The assessment methods which are commonly used in practice are based on principles of the bi-valued logic.

The calculation of the mean value of the scores achieved by each one of its members is the classical method for assessing the ***mean performance*** of a group of objects (e.g. students, players, machines, etc.) with respect to an action.

On the other hand, a very popular in the USA and other Western countries assessment method is the calculation of the ***Grade Point Average (GPA) index***. This index is a weighted average in which greater coefficients (weights) are assigned to the higher scores. GPA, which is connected to the ***quality group's performance*** is calculated by the formula

$$\text{GPA} = \frac{0n_F + 1n_D + 2n_C + 3n_B + 4n_A}{n} \quad (1).$$

In the above formula n is the total number of the group's members and n_A, n_B, n_C, n_D and n_F denote the numbers of those members that demonstrated excellent (A), very good (B), good (C), fair (D) and unsatisfactory (F) performance respectively (Swinburne.edu, 2014)

.

In case of the worst performance $(n_F = n)$ formula (1) gives that GPA = 0, while in case of the ideal performance $(n_A = n)$ it gives GPA = 4. Therefore we have in general that $0 \leq \text{GPA} \leq 4$. Consequently, values of GPA greater than 2 indicate a more than satisfactory performance.

THE COG DEFUZZIFICATION TECHNIQUE: AN OVERVIEW

As we have already seen before, the last step of a system's study using FL techniques is the defuzzification of the previously obtained fuzzy data in order to implement the solution found to the original, real-world problem. The ***Centre of Gravity (COG) Technique*** is the most popular defuzzification method of FL. The development of this method is sketched below:

Let us assume that the system's data are represented by the fuzzy set

$A = \{(x, m(x)): x \in U\}$ on U. We correspond to each $x \in U$ an interval of values from a prefixed numerical distribution, which actually means that we replace U with a set of real intervals. Then, we construct the graph F of the membership function $y=m(x)$. There is a commonly used in FL approach (e.g. van Broekhoven & De Baets, 2006) to represent the system's fuzzy data with the coordinates (x_c, y_c) of the COG, say F_c, of the graph F, which we calculate using the following well-known from Mechanics (Wikipedia.org, 2014a) formulas:

$$x_c = \frac{\iint_F x\,dx\,dy}{\iint_F dx\,dy},\quad y_c = \frac{\iint_F y\,dx\,dy}{\iint_F dx\,dy} \qquad (2).$$

RECTANGULAR FUZZY ASSESSMENT MODEL (RFAM)

As we have already mentioned before, situations appear frequently in our day to day life characterized by a degree of vagueness and/or uncertainty. When an assessment is needed in such kind of situations, a crisp characterization is not usually the proper one. A teacher, for example, is frequently not sure about the exact numerical grade characterizing a student's performance.

FL, due to its property of characterizing the ambiguous cases with multiple values, offers rich resources for this kind of assessment situations. The measurement of a fuzzy system's uncertainty, presented in Chapter 5, provides an example of assessment under fuzzy conditions.

Subbotin et al. (2004), based on Voskoglou's (1999) fuzzy model for learning, which we have developed in Section 4.6, adapted properly the COG technique to be used as a student assessment method. In the following years Voskoglou and Subbotin working either jointly or independently have applied the COG method for assessing many other human and machine activities (see Section 6.11). A detailed description follows here of the general framework that they have used for their purposes:

Let U = {A, B, C, D, F} be the set of the ***fuzzy linguistic labels*** of excellent (A), very good (B), good (C), fair (D) and unsatisfactory (F) performance respectively, of a group of people or of any other objects. When a score, say y, is assigned to a group's member (e.g. a mark in case of a student), then its performance is characterized by F, if y $\in$ [0, 1) , by D, if y $\in$ [1, 2), by C, if y $\in$ [2, 3), by B if y $\in$ [3, 4) by A if y $\in$ [4,5] respectively. Consequently, we have that $y_1 = m(x) = m(F)$ for all x in [0,1), $y_2 = m(x) = m(D)$ for all x in [1,2), $y_3 = m(x) = m(C)$ for all x in [2, 3), $y_4 = m(x) = m(B)$ for all x in [3, 4) and $y_5 = m(x) = m(A)$ for all x in [4,5]. Then the graph of the membership function $y = m(x)$, takes the form of Figure 1, where the area of the level's section S contained between the graph and the OX axis is equal to the sum of the areas of the rectangles S_i, i=1, 2, 3, 4, 5.

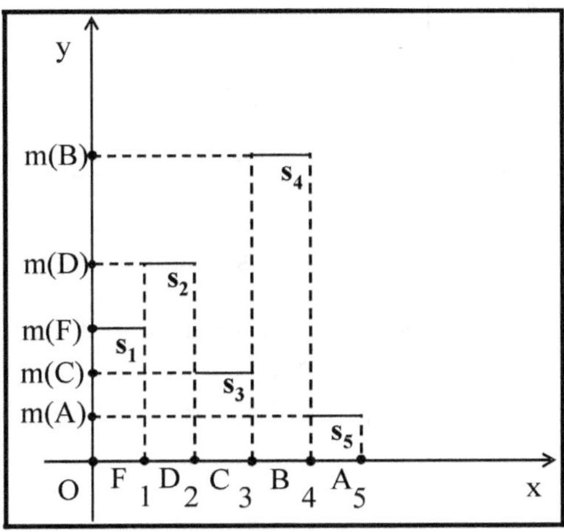

Figure 1: The graph of the COG method

It is straightforward to check that in this case formulas (2) take the form:

$$x_c = \frac{1}{2}\left(\frac{y_1 + 3y_2 + 5y_3 + 7y_4 + 9y_5}{y_1 + y_2 + y_3 + y_4 + y_5}\right),\quad y_c = \frac{1}{2}\left(\frac{y_1^2 + y_2^2 + y_3^2 + y_4^2 + y_5^2}{y_1 + y_2 + y_3 + y_4 + y_5}\right) \quad (3),$$

with x_1=F, x_2=D, x_3=C, x_4=B and x_5=A and $y_i = m(x_i)$, i=1, 2, 3, 4, 5.

In fact, $\iint_S dxdy$ is the area of S which is equal to $\sum_{i=1}^{5} y_i$. Also $\iint_S xdxdy = \sum_{i=1}^{5} \iint_{S_i} xdxdy = \sum_{i=1}^{5} \int_0^{y_i} dy \int_{i-1}^{i} xdx$

$$=\sum_{i=1}^{5} y_i \int_{i-1}^{i} x dx = \sum_{i=1}^{5} y_i [\frac{x^2}{2}]_{i-1}^{i} = \frac{1}{2}\sum_{i=1}^{5} y_i [i^2-(i-1)^2] = \frac{1}{2}\sum_{i=1}^{5}(2i-1)y_i,$$

while $\iint_S ydxdy = \sum_{i=1}^{5}\iint_{F_i} ydxdy = \sum_{i=1}^{5}\int_0^{y_i} ydy \int_{i-1}^{i} dx = \sum_{i=1}^{n}\int_0^{y_i} ydy = \frac{1}{2}\sum_{i=1}^{n} y_i^2$. The result follows by replacing the above values to formulas (2).

Normalizing the membership degrees by dividing each y_i by the sum $\sum_{i=1}^{5} y_i$ we may assume without loss of generality that $\sum_{i=1}^{5} y_i = 1$. Therefore formulas (3) can be finally written in the form:

$$x_c = \frac{1}{2}(y_1+3y_2+5y_3+7y_4+9y_5),\quad y_c = \frac{1}{2}(y_1^2+y_2^2+y_3^2+y_4^2+y_5^2) \qquad (4).$$

In the above equation x_1=F, x_2=D, x_3=C, x_4=B, x_5=A and $y_i = \frac{m(x_i)}{\sum_{j=1}^{5} m(x_j)}$, i = 1, 2, 3, 4, 5.

But $0 \leq (y_i - y_j)^2 = y_i^2 + y_j^2 - 2y_iy_j$, or $y_i^2 + y_j^2 \geq 2y_iy_j$ for i, j = 1, 2, 3, 4, 5, with the equality holding if, and only if, $y_i = y_j$. Adding all the above inequalities by members one finds that $4\sum_{i=1}^{5} y_i^2 \geq 2\sum_{\substack{i,j=1\\ i\neq j}}^{5} y_iy_j$, or $5\sum_{i=1}^{5} y_i^2 \geq 2\sum_{\substack{i,j=1\\ i\neq j}}^{5} y_iy_j + \sum_{i=1}^{5} y_i^2 = (\sum_{i=1}^{5} y_i)^2$, i.e. that $(y_1+y_2+y_3+y_4+y_5)^2 \leq 5(y_1^2+y_2^2+y_3^2+y_4^2+y_5^2)$, with the equality holding if, and only if $y_1 = y_2 = y_3 = y_4 = y_5$. But $y_1 + y_2 + y_3 +$

$y_4 + y_5 = 1$, therefore $1 \leq 5(y_1^2+y_2^2+y_3^2+y_4^2+y_5^2)$ (5), with the equality holding if and only if $y_1 = y_2 = y_3 = y_4 = y_5 = \frac{1}{5}$.

In the last case the first of formulas (4) gives that $x_c = \frac{5}{2}$. Further, combining the inequality (5) with the second of formulas (4), one finds that $1 \leq 10y_c$, or $y_c \geq \frac{1}{10}$. Therefore the unique minimum for y_c corresponds to the COG $F_m(\frac{5}{2}, \frac{1}{10})$.

The ideal case is when $y_1 = y_2 = y_3 = y_4 = 0$ and $y_5=1$. Then from formulas (4) one gets that $x_c = \frac{9}{2}$ and $y_c = \frac{1}{2}$. Therefore the COG in this case is the point $F_i(\frac{9}{2}, \frac{1}{2})$. On the other hand, in the worst case $y_1=1$ and $y_2 = y_3 = y_4 = y_5 = 0$. Then using formulas (4), one finds that the COG is the point $F_w(\frac{1}{2}, \frac{1}{2})$. Thus, the area in which the COG F_c lies is represented by the triangle $F_w F_m F_i$ of Figure 2.

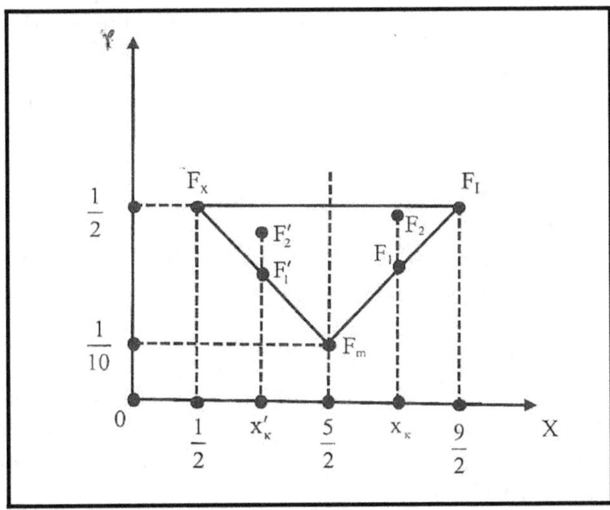

Figure 2: Graphical representation of the area in which the COG lies

Then from elementary geometric considerations it follows that the greater is the value of x_c the better is the group's performance. Also, for two groups with the same $x_c \geq 2.5$, the group having the COG which is

situated closer to F_i is the group with the higher y_c, while for two groups with the same $x_c < 2.5$ the group having the COG which is situated farther to F_w is the group with the lower y_c. Based on the above considerations it is logical to formulate our criterion for comparing the groups' performances in the following form:

- *Among two or more groups the group with the biggest x_c performs better.*
- *If two or more groups have the same $x_c \geq 2.5$, then the group with the higher y_c performs better.*
- *If two or more groups have the same $x_c < 2.5$, then the group with the lower y_c performs better.*

As it becomes evident from the above statement, when applying the COG technique as an assessment method, a group's performance depends mainly on the value of the x-coordinate of the COG of the corresponding graph, which is calculated by the first of formulas (4). In this formula greater coefficients (weights) are assigned to the higher grades. Therefore, the COG method focuses, similarly to the GPA index, on the group's ***quality performance.*** Further, since the value $\frac{9}{2}$ of the COG's x-coordinate corresponds to the group's ideal performance, values of the x-coordinate greater than half of the above value, i.e. greater than $\frac{9}{4}$, demonstrate a more than satisfactory group's performance.

As we have seen above, when the COG defuzzification technique is adapted for use as an assessment method, its scheme consists of five adjacent rectangles with one of their sides lying on the X - axis (Figure 14). For this reason the method was named as the ***Rectangular Fuzzy Assessment Model (RFAM)***.

ASSESSING THE RESULTS OF ITERATIVE LEARNING

The Iterative Instructional Model for Learning

It is recalled that iteration is the repeated application of a procedure, where each step is applied to the output of the preceding one (Borowski & Borwein, 1991). In contrast to the traditional consecutive translation along the material with polishing of all details before reaching the next step, the ***Iterative Instructional Model*** (IIM), proposed as an effective approach to improve the efficiency of learning, suggests a holistic procedure exploring all sides of a problem (Bruner, 1960, Greiner et al., 2004, Opperman & Thomas, 2004, etc.). The physiological effectiveness of the iterative approach is based on our memory properties. Our brain codes a learned information and stores it coded (Voss's categorization of knowledge, see Section 1.5). To retrieve this information the brain need to decode it. The speed and efficiency of decoding (retrieving) depend on the amount of ways (connections) of the information decoding (Halpern & Hakel, 2004). The right way of learning requires creating and developing as many ways of decoding as possible. We can reach this goal by practicing, active discussions, and other activities, using all possible kinds of memory (visual, motor, audio, and so on), employing the holistic approach, raising vertical and horizontal connections between main parts and details of the material. One of the very important components is the developing of strong connections with the already learned material.

The main principles of the iterative learning are the following:

1 ***A holistic approach*:** At each stage the instructor considers the whole theme, from the beginning to the end, as one whole thing.

2 ***An expanding pace*:** Every stage is based on its predecessor and brings a new level of understanding and clarity by adding more details and connections.

3 ***A multi-repetitive character*:** Each stage requires repeating of the whole material of the theme. The final stage repeats all details and connections.

4 ***A uniform level of knowledge acquisition:*** Each stage covers the same level of knowledge about the material.

The main benefits of the iterative instructional approach are a high uniform level of learning, a significantly reduced level of the subject anxiety, an easy and fast decoding of the information (retention) and a complete understanding of the theme.

The Classroom Experiment

Subbotin et al. (2006) presented an example from elementary Algebra (solution of quadratic equations in one variable) to illustrate the use of the iterative approach for teaching mathematics to high school students. The method applied in this example was used recently for teaching the quadratic equations to a group of 38 students of the first year class (15-16 years old) in an upper high school (Lyceum) of the city of Patras, Greece (experimental group). In parallel, the same subject was taught in the traditional way by the same teacher to the second group (29 students) of the same class (control group). Note that the two groups were more or less equivalent with respect to their student mathematical skills.

The teaching procedure for the experimental group involved the following steps:

First iteration: The key question is WHAT?
The main goal here is to introduce students to the main topics of the theme, main relations between them and connections to the previously learned materials. We should not worry about very surface student understanding on this stage. The way of realizing is survey, lecture, reading, and so on.

At this stage the instructor after presenting the necessary definitions reminds what is the solution of an equation in general and introduces students to the brief history of equations, including the equations of degree higher than two.

Now we immediately jump almost to the end, namely teach the quadratic formula $\boxed{x = \frac{-b \pm \sqrt{D}}{2a}}$, where $D = b^2 - 4ac$, and its implementations.

Considering examples the instructor discusses the possible number of real solutions. Summarizing, the instructor mentions that the value of the discriminant D (positive, zero or negative) plays crucial role in the problem of determining the number of the real solutions of a quadratic equation.

It would be very appropriate to tell to the students that we already reached the final front of the theme and there will be no more complications. It will reduce their anxiety and build students' self-confidence. For the same reasons it is very important to make sure that everybody in the class on this stage can apply the quadratic formula for solving equations. At this stage we do not need to worry about the logical understanding of how we come to this formula.

Second iteration: The key question is HOW?

Focus on the structure and main details of the theme studied at the first step, their connections, exploring some second line details and relations, their functioning in the whole system, making this system working. On this stage we start to build the infrastructural system of the material. The best ways of realizing are active discussions, projects, solving easy problems. The discussion method helps the students to elaborate the key concepts. This method is especially useful in helping students to learn to evaluate the logic of, and evidence for, their own and others' position and gives them the opportunities to formulate applications of principles.

At this stage it is appropriate to introduce the solution of equations by using factoring and using the formula and to discuss the difference between these methods. At the end every student is supposed to be able to solve any equations using the formula. Moreover, everybody is supposed to be able to explain HOW this formula works, and HOW to apply the factoring method.

***Third iteration**:* The key question is WHY?

Observe the relations of each main part with the system as a whole and the role which this part plays in the process. Trying to select and learn the main reasons for the existence and functioning of each part, exploring detours and shortcuts between parts. On this stage we develop the infrastructure system of the theme. Best ways to realize it are team projects, working in groups, practicing with the problems requiring non-ordinal approach. For example, WHY we cannot factor every trinomial?

After one completes the review we come to the questions WHY the quadratic formula works? In other words, WHY the solutions could be always found in this way?

It is very appropriate to work now on the completing to the square methods. This method naturally leads to the next step -- the quadratic formula we already studied. At this stage we introduce students to the most complicated parts of the theme. But they are already well prepared for this and have a significant experience to work with the content.

Forth iteration: The key question is WHY NOT?

Attention to details. Work neat with each small part. Try to change the order of particles in the maze answering on the question: what happen to the system if we will make this or that permutation? At this stage we seek for the complete understanding of the material. Accompany activities: high-level problems solving requiring creativity and independent research. The best way to accomplish is whole class work, discussion of each problem, special projects and problems requiring whole material involving. A complete review performed by students is very appropriate. It is very suitable here to explain how to factor any trinomial knowing the solutions of the corresponding equations.

***Fifth iteration**:* Assessment.

Iteration is very useful not only in instruction, but in assessment as well. The levels of the iteration process here involve:

1 Iteration of the whole process as a chain of steps of assessments.

2 Iterative assessment of a specific theme acquisition.

3 Iterative structure of the specific test.

When the teaching procedure was completed a common written test was delivered to both groups for assessing the results of the instruction to the theme. The test involved the following types of questions divided in three parts corresponding to the Voss's stages of learning :

PART 1: Knowledge interpretation.

1. Circle the equations among the following expressions.

2. What is the root of an equation?

3. How many solutions does a quadratic equation have?

4. What ways of solving quadratic equations do you know?

5. Write the formula for the solution of a quadratic equation.

PART 2: Knowledge generalization.

6. Describe in your own words how to solve a quadratic equation.

7. Solve a specific given quadratic equation.

8. Graphical interpretation.

PART 3: Knowledge categorization.

9. Application to a specific word problem.

10. Solving a problem with analyzing parameters.

The student papers were marked in a climax from 0 to 100 and the students of the two groups obtained the following scores:

Experimental group (G_1): 99(1 student), 83(2), 82(1), 74(10), 72(2), 70(1), 59(10), 55(2), 48(7), 45(2)
Control group (G_2)***:*** 85(2), 75(1), 62(2), 60(10), 52(1), 50(8), 25(4), 10(1).

We assigned to the linguistic labels (grades) of U the student scores as follows: A (85-100), B(75-84), C(60-74), D(50-59), F(0-49). Note that the above correspondence, although it satisfies the common sense, cannot be considered as a standard one; for example in a more strict assessment one could take A (90-100), B(80-89), C(70-79), D(60-69), F(0-59), etc.

The evaluation of the above results was performed as follows:

i) Calculation of the mean values: A straightforward calculation gives that the mean values of the students' scores are approximately equal to 62.231 for G_1 and to 52.793 for G_2 respectively. This shows that the students of G_1 demonstrated a good (C), while the students of G_2 demonstrated a fair (D) mean performance.

ii) Calculation of the GPA index: The results of the student performance are depicted in Table 1 below:

Table 1: The results of the student performance

Grade	G_1	G_2
A	1	2
B	3	1
C	13	12
D	12	9
F	9	5
Total	38	29

Replacing the data of Table 6 to formula (1) of Section 6.1 one finds that the GPA index is equal to $\frac{51}{38} \approx 1.342$ for G_1 and equal to $\frac{44}{29} \approx 1.517$ for G_2. Therefore, both Departments demonstrated a less than satisfactory quality performance. However, in contrast to their mean performance, the quality performance of G_2 was better than the corresponding performance of G_1.

iii) Application of the RFAM: From Table 1 one finds for G_1 that $y_1 = \frac{9}{38}$, $y_2 = \frac{12}{38}$, $y_3 = \frac{13}{38}$, $y_4 = \frac{3}{38}$ and $y_5 = \frac{1}{38}$. Replacing these values to the first of formulas (4) of Section 6.3 one finds that $x_c = \frac{140}{76} \approx 1.842$. Similarly it turns out that $x_c = \frac{117}{58} \approx 2.017$ for G_2. Since in both cases the value of x_c is less than $\frac{9}{4}$ (see Section 6.3), the two groups demonstrated a less than satisfactory quality performance, while G_2 demonstrated a better performance than G_1. Therefore, the conclusions obtained by applying the RFAM are the same with those of the GPA index.

Based on the above assessment outcomes one concludes that the application of the IIM had a positive effect on the mean performance, but not on the quality performance of the experimental group. This means that only the mediocre students (lower scores) were benefited by the iterative approach of learning, which was logically expected.

BRIDGE PLAYER ASSESSMENT

The Game of Contract Bridge

Contract Bridge is a trick-taking card game which, together with Chess, are the only mind sports officially recognized by the International Olympic Committee. The object is to make at least as many tricks as were contracted for and penalties are imposed for failing to do so. Millions of people play nowadays Bridge worldwide, not only in clubs, tournaments and championships, but also on line and with friends at home, making it one of the world's most popular card games. The *World Bridge Federation (WBF)* is the international governing body of Contract Bridge. WBF was formatted in August 1958 by delegates from Europe, North and South America and its membership today comprises 123 National Bridge Organizations.

A match of Bridge is played either among *teams* (two or more) of four players (two partnerships) in each team, or among *pairs*. For a pairs event a minimum of three tables (6 pairs, 12 players) is needed, but it works better with more players. The usual method of scoring in a pairs' competition is in *Match Points (MPs)*. Each pair is awarded two MPs for each pair who scored worse on each game's session (hand), and one MP for each pair who scored equally. The total number of MPs scored by each pair over all the hands played is calculated and it is converted to a percentage. The pair succeeding the highest percentage wins the game. On the other hand, in a team event the result is the difference in *International Match Points (IMPs)* between the competing teams and then there is a further conversion, in which some fixed number of *Victory Points (VPs)* is appointed between the teams.

For the fundamentals and the rules of Contract Bridge, as well as for the conventions usually played between the partners we refer to the famous book of Edgar Kaplan (1925-1997), who was an American Bridge player and one of the principal contributors to the game. Kaplan's book was translated in many languages and was reprinted many times since its first edition in 1964.There are also many other books written about Bridge as well as a fair amount of related information on the Internet, e.g. see Pagat.com, 2014, etc.

Application of the RFAM to Assessment of Bridge Player Skills

Apart from the official scoring methods of Bridge mentioned in Section 6.5.1, it is useful sometimes for statistical or other reasons to assess the overall performance of certain sets of single players, pairs, or teams. For example, this happens, when one wants to compare the performance of two or more clubs participating in the same tournament, the performance of male and female or of old and young players, etc. This is usually done by calculating the mean values of the official scores obtained.

Voskoglou (2014b) utilized the COG technique as an assessment method for Bridge. Here, we shall present one of the examples appearing in this paper. For this, note that in a pair competition with MPs as the scoring method and according to the usual standards of Bridge, one could characterize performance, according to the percentage of success, say p, as follows:

- Excellent (A), if $p > 65\%$.
- Very good (B), if $55\% < p \leq 65\%$.
- Good (C), if $48\% < p \leq 55\%$.
- Fair (D), if $40\% \leq p \leq 48\%$.
- Unsatisfactory (F), if $p < 40\ \%$.

The above characterization could not be considered as being a standard one, since performance depends on the opponents' skills and on several other special factors appearing in each game.

Our example is related to the total scoring table of the players of a Bridge club from the city of Patras, who participated in at least five of the six in total events of a simultaneous tournament organized by the *Hellenic Bridge Federation*, which ended on February 19, 2014. Nine men and five women players are included in this table, who obtained the following mean scores (the worst score of each player participating to all the tournament's events was dropped out):

Men: 57.22%, 54.77%, 54.77%, 54.35%, 54.08%, 50.82 %, 50.82%, 49.61%, 47.82%.

Women: 59.48%, 54.08%, 53.45%, 53.45%, 47.39%.

The above results give a mean percentage of approximately 52.696% for the men and 53.57% for the women players. Therefore the women demonstrated a slightly better mean performance than the men players.

The player results are depicted in Table 2, according to the above introduced performance characterizations.

Replacing the values of the last column of Table 2 to the first of formulas (4) of Section 6.3 one finds that $x_c = \frac{1}{2}(3.\frac{1}{9}+5.\frac{7}{9}+7.\frac{1}{9}) = \frac{45}{18} = 2.5$ for the men players, and $x_c = \frac{1}{2}(3.\frac{1}{5}+5.\frac{3}{5}+7.\frac{1}{5}) = \frac{25}{10} = 2.5$ for the women players. Further, the second of formulas (4) gives that y_c = $\frac{1}{2}[(\frac{1}{9})^2+(\frac{7}{9})^2+(\frac{1}{9})^2] = \frac{51}{162}$ □ 0.315 for the men and y_c = $\frac{1}{2}[(\frac{1}{5})^2+(\frac{3}{5})^2+(\frac{1}{5})^2] = \frac{11}{50} = 0.22$ for the women players. Thus, according to the second case of the assessment criterion presented aboveand in contrast to the mean performance, the men demonstrated a higher quality performance than the women players.

Table 2: Mean player scoring

Men

% Scale	Level of Performance	Amount of players	Frequency
>65%	A	0	0
55-65%	B	1	1/9
48-55%	C	7	7/9
40-48%	D	1	1/9
<40%	F	0	0
Total		9	

Women

>65%	A	0	0
55-65%	B	1	1/5
48-55%	C	3	3/5
40-48%	D	1	1/5
<40%	F	0	0
Total		5	

Note that, replacing the data of Table 2 to formula (1) one finds that the GPA index is equal to 2 for both men and women players, which means that they demonstrate the same quality performance. The different conclusions obtained by applying the COG and GPA methods for the player assessment , which is logical to happen in marginal cases like the above, are due to the different philosophy of the two methods; GPA is based on the principles of the bi-valued logic, while COG is based on FL. More specifically, writing formula (1) in terms of the frequencies it takes the form GPA = $y_2 + 2y_3 + 3y4 + 4y_5$ and comparing it to the first of formulas (4) it turns out that the COG method is *more sensitive to the higher scores* by assigning greater weights to them than the GPA index.

GENERALIZED RECTANGULAR FUZZY ASSESSMENT MODEL (GRFAM)

Ambiguous assessment cases frequently appear in practice, being at the boundaries between two successive assessment grades; e.g. something like 84-85%, being at the boundaries between A and B. In an effort to treat better such kind of cases, Subbotin (2015) "moved" the rectangles of Figure 14 of Section 6.3 to the left, so that to share common parts (see Figure 16). In this way, the ambiguous cases, being at the common rectangle parts, belong to both of the successive grades. This means that the common rectangle parts must be considered ***twice*** in the corresponding calculations.

The graph of the resulting fuzzy set is now the bold line of Figure 3. However, the method used in the case of RFAM for calculating the coordinates of the COG of the area included between the graph and the X-axis is not the proper one here, because in this way the common rectangle parts are calculated only once. The right method for calculating the coordinates of the COG for this case was fully developed by Subbotin &

Voskoglou (2016) and the resulting framework was called the ***Generalized Rectangular Fuzzy Assessment Model (GRFAM).*** The development of GRFAM involves the following steps:

1. Let y_1, y_2 , y_3, y_4 , y_3 be the *frequencies* a group's members who obtained the grades F, D, C, B, A respectively. Then $\sum_{i=1}^{5} y_i$ = 1 (100%).
2. We take the heights of the rectangles in Figure 3 to have lengths equal to the corresponding frequencies. Also, we allow the sides of the adjacent rectangles lying on the OX axis to share common parts with length equal to the 30% of their lengths, i.e. 0.3 units.

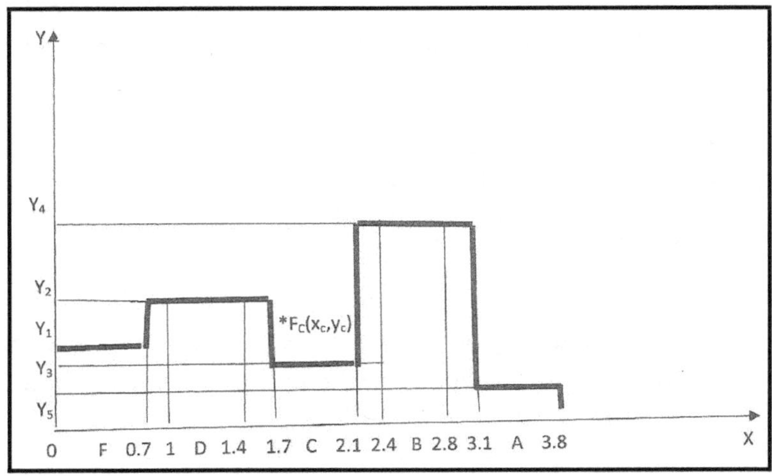

Figure 3: Graphical representation of the GRFAM

3. We calculate the coordinates (x_{c_i}, y_{c_i}) of the COG, say F_i , of each rectangle, i = 1, 2, 3, 4, 5 as follows: Since the COG of a rectangle is the point of the intersection of its diagonals, we have that $y_{c_i} = \frac{1}{2} y_i$. Also, since the x-coordinate of each COG F_i is equal to the x- coordinate of the middle of the side of the corresponding rectangle lying on the OX axis, from Figure 16 it is easy to observe that x_{c_i} = 0.7i – 0.2.

4. We consider the system of the COGs F_i and we calculate the coordinates (X_c, Y_c) of the COG F of the whole area considered in Figure 11 as the resultant of the system of the GOCs F_i of the five rectangles from

the following well known (Wikipedia.org, 2014c) formulas $X_c = \frac{1}{S}\sum_{i=1}^{5} S_i x_{c_i}$, $Y_c = \frac{1}{S}\sum_{i=1}^{5} S_i y_{c_i}$ (6).

In the above formulas *Si,* i= 1, 2, 3, 4, 5 denote the areas of the corresponding rectangles, which are equal to y_i . Therefore $S = \sum_{i=1}^{5} S_i = \sum_{i=1}^{5} y_i = 1$ and formulas (6) give that $X_c = \sum_{i=1}^{5} y_i(0.7i-0.2)$, $Y_c = \sum_{i=1}^{5} y_i(\frac{1}{2}y_i)$ or

$X_c = (0.7\sum_{i=1}^{5} iy_i)-0.2$, $Y_c = \frac{1}{2}\sum_{i=1}^{5} y_i^2$ (7).

5. We determine the area in which the COG F lies as follows: For i, j = 1, 2, 3, 4, 5, we have that $0 \leq (y_i - y_j)^2 = y_i^2 + y_j^2 - 2y_iy_j$, therefore $y_i^2 + y_j^2 \geq 2y_iy_j$, with the equality holding if, and only if, $y_i = y_j$. Therefore $1 = (\sum_{i=1}^{5} y_i)^2 =$

$= \sum_{i=1}^{5} y_i^2 + 2\sum_{i,j=1,\, i\neq j}^{5} y_iy_j \leq \sum_{i=1}^{5} y_i^2 + 2\sum_{i,j=1,\, i\neq j}^{5} (y_i^2+y_j^2) = 5\sum_{i=1}^{5} y_i^2$ or $\sum_{i=1}^{5} y_i^2 \geq \frac{1}{5}$ (8), with the equality holding if, and only if, $y_1 = y_2 = y_3 = y_4 = y_5 = \frac{1}{5}$. In case of the equality the first of formulas (7) gives that $X_c = 0.7(\frac{1}{5} + \frac{2}{5} + \frac{3}{5} + \frac{4}{5} + \frac{5}{5}) - 2 = 1.9$. Further, combining the inequality (8) with the second of formulas (7), one finds that $Y_c \geq \frac{1}{10}$ Therefore the *unique minimum for* Y_c corresponds to the COG

$F_m(1.9,\ 0.1)$.

The *ideal case* is when $y_1 = y_2 = y_3 = y_4 = 0$ and $y_5=1$. Then formulas (2) give that $X_c = 3.3$ and $Y_c = \frac{1}{2}$. Therefore the COG in this case is the point

$F_i(3.3, 0.5)$.

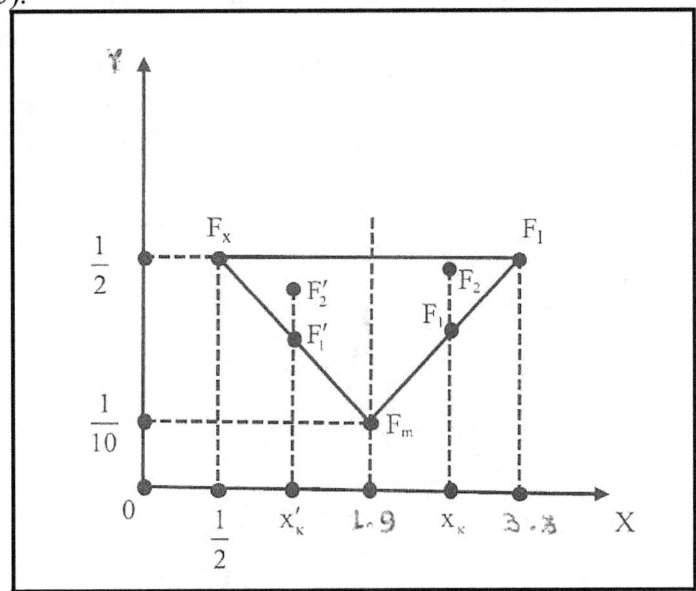

Figure 4: The triangle where the COG lies

On the other hand, the *worst case* is when $y_1 = 1$ and $y_2 = y_3 = y_4 = y_5 = 0$. Then from formulas (2) we find that the COG is the point F_w (0.5, 0.5). Therefore, the area in which the COG F lies is the area of the triangle $F_w F_m F_i$ (Figure 4).

6. From elementary geometric observations on Figure 4, analogous to those performed on Figure 15 of Section 6.3, one obtains the following assessment criterion:

- *Between two groups, the group with the greater X_c performs better.*
- *If two groups have the same $X_c \geq 1.9$, then the group with the greater Y_c performs better*
- *If two groups have the same $X_c < 1.9$, then the group with the lower Y_c performs better.*

TRIANGULAR FUZZY ASSESSMENT MODEL (TFAM)

At this point one could ask the following question: Does the shape of the membership function's graph in the assessment model affect the assessment's conclusions? For example, what will happen if the rectangles of the GRFAM are replaced by triangles? The effort to answer this question leads to the construction of the ***Triangular Fuzzy Assessment Model (TFAM)***, created by Subbotin & Bilotskii (2014) and fully developed by Subbotin & Voskoglou (2014b). The headlines of TFAM are the following:

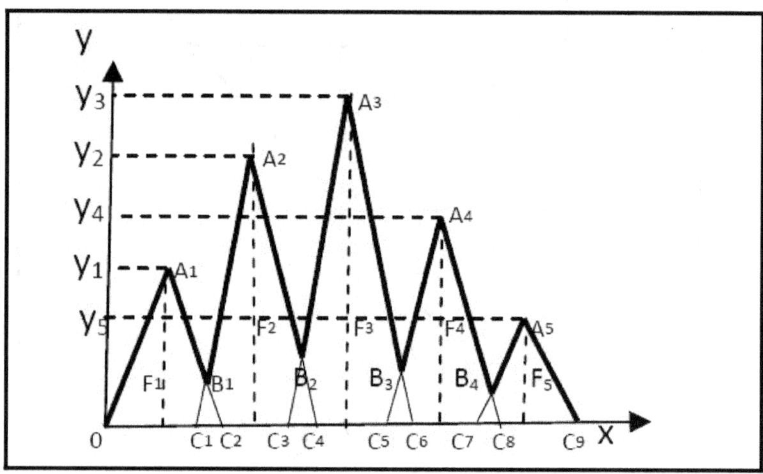

Figure 5: Graphical Representation of the TFAM

- We consider the isosceles triangles with bases having lengths of one unit and their heights to the base equal to the frequencies y_1, y_2, y_3, y_4, y_5 (see Figure 18). Each pair of adjacent triangles has common parts on the base of length 0.3 units.

- We calculate the coordinates (x_{c_i}, y_{c_i}) of the COG F_i, i = 1, 2, 3, 4, 5 of each triangle, which is the point of the intersection of its medians. Since F_i divides the median to proportion 2:1 from the vertex and the triangles are isosceles, we find that $y_{c_i} = \frac{1}{3} y_i$. Also, since the x-coordinate of each COG F_i is equal to the x- coordinate of the middle of the side of the corresponding triangle lying on the OX axis, from Figure 5 it is easy to observe that x_{c_i} = 0.7i – 0.2.

- We consider the system of the COGs F_i, i=1, 2, 3, 4, 5 and we calculate the coordinates (X_c, Y_c) of the COG F_c of the whole level's area considered in Figure 5 by formulas (6), where Si = $0.5y_i$ and S = $\sum_{i=1}^{5} S_i$ = 0.5$\sum_{i=1}^{5} y_i$ = 0.5. Thus, one finds that the coordinates of the COG of the resulting in this case scheme are calculated by the formulas

$$X_c = (0.7\sum_{i=1}^{5} iy_i) - 0.2, \quad Y_c = \frac{1}{3}\sum_{i=1}^{5} {y_i}^2 \quad (9).$$

- Working as in the paragraphs 5 and 6 of Section for GRFAM we obtain the same criterion for comparing the performance of two (or more) groups.

As it can be easily observed from formulas (7), the assessment criterion of Section 6.6 and from formulas (9) the only difference between the outcomes of the GRFAM and the TFAM concerns the value of the coordinate Y_c of the corresponding COG, but this does not affect the assessment conclusions. Therefore, using the one or the other model makes no difference.

TRAPEZOIDAL FUZZY ASSESSMENT MODEL (TPFAM)

An equivalent to the TFAM approach is to consider isosceles trapezoids instead of triangles (Subbotin, 2014, Subbotin and Voskoglou 2014c). In this case we called the resulting framework ***Trapezoidal Fuzzy Assessment Model (TpFAM)***. The corresponding scheme is that shown in Figure 6.

In this case the coordinates (x_{c_i}, y_{c_i}) of the COG F_i, i=1, 2, 3, 4, 5, of each trapezoid are calculated as follows: It is well known that the COG of a trapezoid lies on the line segment joining the midpoints of its parallel sides *a* and *b* at a distance *d* from the longer side *b* given by $d=\frac{h(2a+b)}{3(a+b)}$, where h is its height (Wikipedia.org 2014b).Therefore, here we have y_{c_i} = $\frac{y_i(2*4+10)}{3*(4+10)}=\frac{3y_i}{7}$. Also, since the x-coordinate of the COG of each trapezoid is equal to the x-coordinate of the midpoint of its base, it is easy to observe that x_{ci} = 0.7i - 0.2.

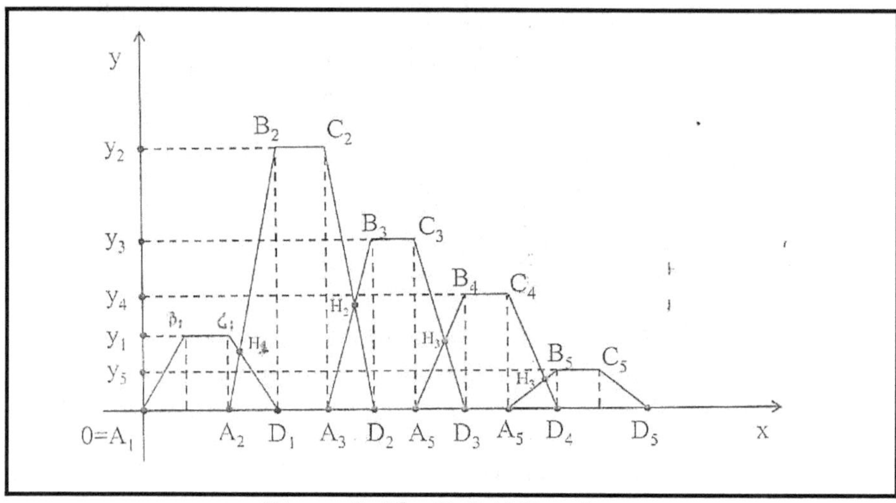

Figure 6: The TpFAM's scheme

We calculate the coordinates (X_c, Y_c) of the COG F_c of the whole level's area considered in Figure 19 by formulas (6) of Section 6.7, where *Si,* i= 1, 2, 3, 4, 5 denotes the area of the corresponding trapezoid. Therefore, *Si* = $\frac{(1+0.4)y_i}{2}$ = $0.7y_i$ and S = $\sum_{i=1}^{5} S_i$ = $0.7\sum_{i=1}^{5} y_i$ = 0.7. Consequently, one finally obtains from formulas (6) that X_c = $(0.7\sum_{i=1}^{5} iy_i)-0.2$, $Y_c=\frac{3}{7}\sum_{i=1}^{5} y_i^2$ (10).

COMPARISON OF THE ASSESSMENT MODELS

One can write formulas (7), (9) and (10) in the single form:

$$X_c=(0.7\sum_{i=1}^{5} iy_i)-0.2,\ Y_c=a\sum_{i=1}^{5} y_i^2 \qquad (11).$$

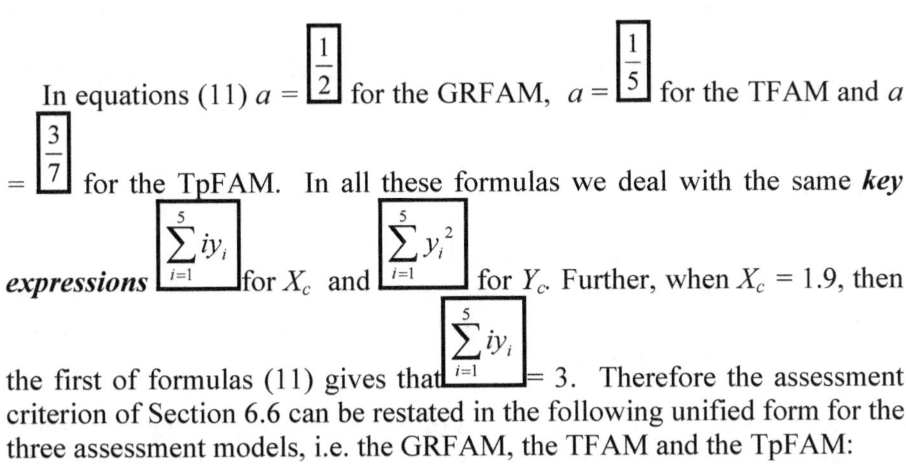

In equations (11) $a = \frac{1}{2}$ for the GRFAM, $a = \frac{1}{5}$ for the TFAM and $a = \frac{3}{7}$ for the TpFAM. In all these formulas we deal with the same ***key expressions*** $\sum_{i=1}^{5} iy_i$ for X_c and $\sum_{i=1}^{5} y_i^2$ for Y_c. Further, when $X_c = 1.9$, then the first of formulas (11) gives that $\sum_{i=1}^{5} iy_i = 3$. Therefore the assessment criterion of Section 6.6 can be restated in the following unified form for the three assessment models, i.e. the GRFAM, the TFAM and the TpFAM:

Between two groups the group with the greater value of $\sum_{i=1}^{5} iy_i$ *performs better.*

If two groups have the same value of $\sum_{i=1}^{5} iy_i$ *then if* $\sum_{i=1}^{5} iy_i \geq 3$, *the group with the greater value of* $\sum_{i=1}^{5} y_i^2$ *performs better, while if* $\sum_{i=1}^{5} iy_i < 3$, *the group with the lower value of* $\sum_{i=1}^{5} y_i^2$ *performs better.*

Combining formulas (11) with the above criterion one obtains the following result:

1. Theorem: *The three variations of the COG technique, i.e. the GRFAM, the TFAM and the TpFAM are equivalent assessment models.*

Further, the first of formulas (11) can be written as $X_c = 0.7(y_1 + 2y_2 + 3y_3 + 4y_4 + 5y_5) - 0.2 = 0.7\,[(y_2 + 2y_3 + 3y_4 + 4y_5) + \sum_{i=1}^{5} y_i] - 0.2$. Therefore, by formula (1) of Section 6.1, one finally gets that $X_c = 0.7(\text{GPA} + 1) - 0.2$, or $X_c = 0.7\text{GPA} + 0.5$ (12).

In the same way, the first of formulas (4) of Section 6.3 for RFAM can be written as $x_c = \frac{1}{2}(y_1 + 3y_2 + 5y_3 + 7y_4 + 9y_5) = \frac{1}{2}(2\text{GPA} + 1)$, or $x_c = \text{GPA} + 0.5$ (13).

We are ready now to prove:

2. Theorem: *If the values of the GPA index are different for two groups, then the GPA index, the RFAM and its variations (GRFAM, TFAM and TpFAM) provide the same assessment outcomes on comparing the performance of these groups.*

Proof: Let G and G′ be the values of the GPA index for the two groups and let x_c, x_c' be the corresponding values of the x-coordinate of the COG for the RFAM. Assume without loss of generality that G>G′, i.e. that the first group performs better according to the GPA index. Then, equation (13) gives that $x_c > x_c'$, which, according to the first case of the assessment criterion of Section 6.3, shows that the first group performs also better according to the RFAM.

In the same way, from equation (12) and the first case of the assessment criterion of Section 6.6, one finds that the first group performs better too according to the equivalent assessment models GRFAM, TFAM and TpFAM.-

In case of the same GPA index we shall show the following result:

3. Theorem: *If the GPA index is the same for two groups then the RFAM and its variations (GRFAM, TFAM and TpFAM) provide the same assessment outcomes on comparing the performance of these groups.*

Proof: Since the two groups possess the same value of the GPA index, equations (12) and (13) show that the values of X_c and x_c are also the same for the two groups. Therefore, one of the last two cases of the assessment criteria of Sections 6.3 and 6.6 could happen. The possible values of x in these criteria lie in the intervals $[0, \frac{9}{2}]$ and [0, 3.3] respectively, while the critical points correspond to the values $x_c = 2.5$ and $X_c = 1.9$ respectively. Obviously, if both values of x are in [0, 1.9), or in $[2.5, \frac{9}{2}]$, then the two criteria provide the same assessment outcomes on comparing the performance of the two groups. Assume therefore that 1.9 < X_c and x_c < 2.5. Then, due to equation (12), 1.9 < X_c $\Leftrightarrow$ 1.9< 0.7GPA + 0.5 $\Leftrightarrow$ 1.4 <0.7GPA $\Leftrightarrow$ GPA > 2.

Also, due to equation (13), x_c < 2.5 $\Leftrightarrow$ GPA + 0.5 < 2.5 $\Leftrightarrow$ GPA < 2. Therefore, the inequalities 1.9 < X_c and x_c < 2.5 cannot hold simultaneously and the result follows.-

Combining Theorems 2 and 3 one obtains the following corollary:

4. Corollary: *The RFAM and its variations GRFAM, TFAM and TpFAM provide always the same assessment results on comparing the performance of two groups*

.

The following example (paragraph (vii) of Section 4 of Subbotin & Voskoglou, 2016) shows that in case of the *same GPA values the application of the GPA index could not lead to logically based conclusions.* Therefore, in such situations, our criteria of Sections 6.3 and 6.6 become useful due to their logical nature.

5. Example: The student grades of two Classes with 60 students in each Class are presented in Table 3

The GPA index for the two classes is equal to $\frac{2*10+4*50}{60}=\frac{3*20+4*40}{60}\approx 3.67$, which means that the two Classes demonstrate the same performance in terms of the GPA index. Therefore equation (12) gives that X_c = 0.7*3.67 + 0.5 ≈ 3.07, while equation (13) gives that x_c = 4.17 for both Classes. But $\sum_{i=1}^{5} y_i^2$ = $(\frac{1}{6})^2+(\frac{5}{6})^2=\frac{26}{36}$ for the first and $\sum_{i=1}^{5} y_i^2$ = $(\frac{2}{6})^2+(\frac{4}{6})^2$ = $\frac{20}{36}$ for the second Class. Therefore, according to the assessment criteria of Sections 6.3 and 6.6 the first Class demonstrates a better performance in terms of the RFAM and its variations.

Table 3: Student Grades

Grades	Class I	Class II
C	10	0
B	0	20
A	50	40

Now which one of the above two conclusions is closer to the reality? For answering this question, let us consider first the *quality of knowledge*, i.e. the ratio of the students received B or better to the total number of students, which is equal to $\frac{5}{6}$ for the first and 1 for the second Class. Therefore, from the common point of view, the situation in Class II is better.

Also, if we assign to the grades A, B, C, D and F the numbers 5, 4, 3, 2 and 1 respectively, we find for Class I the mean values $\overline{X} = \frac{3*10+5*50}{60} \approx$ 4.67 and $\overline{X^2} = \frac{3*10^2+5*50^2}{60} \approx 213.33$. Therefore the variance of X is equal to 213.33 – $(4.67)^2$ $\approx$ 191.52. In the same way one finds that the variance of X for Class II is equal to 160 – $(4.67)^2$ $\approx$138.19 < 213.33. Therefore the standard deviation for the second Class is definitely smaller, which means that, from the statistical point of view, the situation in Class II is also better.

However, many instructors could prefer the situation in Class I having a greater number of excellent students. Conclusively, in no case it is logical to accept that the two Classes demonstrated the same performance, as the calculation of the GPA index suggests.

The next example shows that although the RFAM, GRFAM, TFAM and TpFAM provide always the same assessment results on comparing the performance of two groups (Corollary 4), *they are not equivalent assessment models*

6. Example: Table 4 depicts the results of the final exams of the first term mathematical courses of two different Departments, say D_1 and D_2, of the School of Management and Economics of the Graduate T. E. I. of Western Greece. Note that the contents of the two courses and the instructor were the same for the two Departments.

Table 4: Results of the two Departments

Grade	D_1	D_2
A	1	1
B	3	6
C	11	13
D	9	10
F	6	5
Total No. of students	30	35

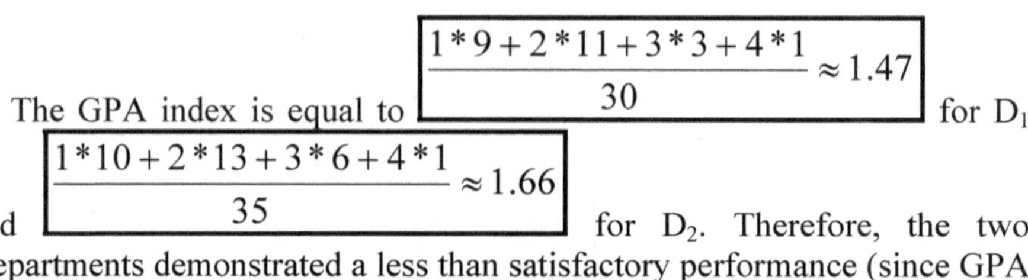

The GPA index is equal to $\frac{1*9+2*11+3*3+4*1}{30} \approx 1.47$ for D_1 and $\frac{1*10+2*13+3*6+4*1}{35} \approx 1.66$ for D_2. Therefore, the two Departments demonstrated a less than satisfactory performance (since GPA < 2), with the performance of D_2 being better.

Further, equation (12) gives that $X_c \approx 1.53$ for D_1 and $X_c \approx 1.66$ for D_2. Therefore, according to the first case of the assessment criterion of Section 6.6, D_2 demonstrated (with respect to GRFAM, TFAM and TpFAM) a better performance than D_1. But, since $1.53 < \frac{3.3}{2} = 1.65 < 1.66$, D_1 demonstrated *a less than satisfactory performance*, while D_2 demonstrated *a more than satisfactory performance.*

In the same way equation (13) gives that $x_c \approx 1.97$ for D_1 and $x_c \approx 2.16$ for D_2. Therefore, according to the first case of the assessment criterion of Section 6.3, D_2 demonstrated (with respect to RFAM) a better performance than D_1. But in this case, since for both Departments $x_c < \frac{4.5}{2} = 2.25$, *the two Departments demonstrated a less than satisfactory performance.*

7. ***Theorem:*** *The assessment conclusions of RFAM and of the equivalent GRFAM, TFAM, TpFAM can be different only when GPA < 2*

.

Proof: If GPA > 2 (more than satisfactory performance), then
$X_c = 0.7\text{GPA} + 0.5 > 0.7 * 2 + 0.5 = 1.9 > 1.65$ and
$x_c = \text{GPA} + 0.5 > 0.2 + 0.5 = 2.5 > 2.25$.

Therefore the corresponding group's performance is also more than satisfactory with respect to GRFAM, TFAM, TpFAM and RFAM.

However, if GPA < 2 (less than satisfactory performance), then $X_c < 1.9$ and
$x_c < 2.5$, which do not guarantee that $X_c < 1.65$ and $x_c < 2.25$.

8. Remark: For the development of GRFAM, TFAM and TpFAM we took the common lengths involved in Figures 17, 18 and 19 respectively to be equal to the 30% of the lengths of their bases. Nevertheless, it is easy to check that the key expressions $\sum_{i=1}^{5} iy_i$ for X_c and $\sum_{i=1}^{5} y_i^2$ for Y_c remain unchanged for percentages k% of the common lengths, with $k \neq 30$. Note however that, since the ambiguous assessment cases are situated at the boundaries between the adjacent grades, which means that their x-coordinates take values near to the edges of the corresponding to these grades real intervals, it is logical to accept that k < 50. Moreover, if $k \geq 50$, then in Figures 17, 18 and 19 the interval of C on the OX axis will be completely covered by the intervals of D and B.

In general, assigning the interval [0, 1] to F and considering our graphs on the interval [0, m] and the bases of the adjacent figures sharing *k*% of their lengths, it is easy to check that m = 5 - $\frac{4k}{100}$. In fact, initially the interval [0, 5] is needed to construct the five rectangles of the RFAM (or triangles of the TFAM, or trapezoids of the TpFAM), while the movement of these figures to the left for sharing common parts reduces the length (5 units) of this interval by $\frac{4k}{100}$ units. It is also possible to consider n assessment grades with $n \neq 5$; in this case we obviously have that m = n - $\frac{(n-1)k}{100}$.

APPLICATION TO ASSESSMENT OF CRITICAL THINKING SKILLS

Critical Thinking: An Overview

Substantial changes took place in the structures of the human Society during the last 150 years. The ***First Industrial Revolution***, started at the end of the 19th Century, was characterized by the replacement of human hands by machines as power sources with a parallel development of transportation and communication means. The first was followed by the ***Second Industrial Revolution***, better known as the era of ***Automatism,*** started during the 1940s' and characterised by the replacement of humans by machines (mainly by the computers) as means of control. As a result, the human Society has been transformed to a ***Society of Knowledge and Globalization***. Note that, nowadays we are in front of the forthcoming era of the ***Third Industrial Revolution*** which, according to the best shelling author J. Rifkin (2016), will be characterized by the "***Internet of Things and Energy***". Through this, energy, goods and services will be provided, almost free of cost, in the same way as information does through our nowadays Web.

All the above changes have obviously made the problems of our day to day life much more composite and complicated than they used to be 150 years ago. Such kind of problems require a higher-order thinking for their solution, Such kind of thinking, usually referred as ***Critical Thinking*** (CrT) involves synthesis and analysis, abstraction, uncertainty, application of multiple criteria, reflection, decision making, drawing warrant conclusions and generalizations, self-regulation, etc. It also facilitates the ***transfer of knowledge***, i.e. the use and transformation of already existing knowledge for creating new knowledge. The complexity of CrT is evident from the fact that there is no definition about it that is universally accepted. Some of the most characteristic definitions existing in the literature are the following: "...disciplined thinking that is clear, rational, open-minded, and informed by evidence" (Dictionary.reference.com, 2013); "the skill and propensity to engage in an activity with reflective skepticism" (Mc Peck, 1992) "...disciplined, self-directed thinking which exemplifies the perfection of thinking appropriate to a particular mode of domain of thinking." (Paul, 1982), etc.

Assessing student CrT skills: A classroom application

Beyond understanding theory and formulas, the students need to be proficient in application of mathematics and science knowledge to different situations and challenges. That is why just the well developed *reading comprehending skills* are crucially important for solving mathematical content problems. In fact, even being skillful in the formal technical mathematics and communication of mathematics, the student, whose reading comprehension abilities are limited, will not be able to make any progress in the application of these mathematical skills to some problems or just simple questions related to real world. Subbotin & Voskoglou (2014a, 2016), in their will to support this important statement, presented the following interesting example:

In one of the Los Angeles Unified District inner city schools having very diverse student population (Hispanic 53% , Asian 22%, Black 18%,

White 7%) the Algebra 2 District Assessment Test was given. The test's questions can be found in next Section. A very professional and dedicated teacher, who conducted this test, gave it in two of his Algebra 2 classes. The one of them was a *regular class*, while the other was a so-called *shelter class*, which means that the waist majority of the students in this class were students for whom English is a second language, not a native tongue. The results of the test are presented in Tables 5 and 6 below .for the shelter and the regular class respectively.

Table 5: Results of the shelter class

% Scale	Grade	Students
85-100	A	0
75-84	B	5
60-74	C	6
50-59	D	9
Less than 50	F	18
Total		38

Table 6: Results of the regular class

% Scale	Grade	Students
85-100	A	0
75-84	B	1
60-74	C	5
50-59	D	3
Less than 50	F	20
Total		29

Next we compare the outcomes obtained for the two classes as follows:

(i) Mean values: Assigning the values 5, 4, 3, 2, 1 to the grades A, B, C, D, F respectively one finds the mean values $\frac{5*4+6*3+9*2+18.1}{38} \approx 1.95$ for the shelter and

$\frac{1*4+5*3+3*2+20*1}{29} \approx 1.55$ for the regular class respectively. Therefore, the shelter class demonstrated a better *mean performance*.

(ii) GPA index: One finds that GPA=$\frac{9*1+6*2+5*3}{38} \approx 0.95$ for the shelter and GPA=$\frac{3*1+5*2+1*3}{29} \approx 0.55$ for the regular class. Since both values of the GPA index are smaller than 2, both classes demonstrated a less than satisfactory *quality performance*; however the shelter class demonstrated again a better performance.

Since the values of the GPA index are different for the two classes, by Theorem 6.9.2 the assessment conclusion on comparing the performance of the two classes that will be obtained by applying the COG technique (RFAM) or one of its equivalent variations (GRFAM, TFAM and TpFAM) will be the same with that obtained by calculating the GPA index.

(iii) *RFAM:* Applying the first of formulas (4) of Section 6.3 one finds that

$x_c = \frac{1}{2} * \frac{18+3*9+5*6+7*5}{38} \approx 1.45$ for the shelter and

$x_c = \frac{1}{2} * \frac{20+3*3+5*5+7*1}{29} \approx 1.05$ for the regular class respectively. Therefore, according to the criterion of Section 6.3 the shelter class demonstrated a better performance. Further, since the value of x_c for both classes is much smaller than the half of its value in the ideal case ($\frac{9}{2}$: 2 = 2.25), their quality performance is characterized as unsatisfactory.

(iv) *GRFAM (or TFAM or TpFAM):* The first of formulas (7) of Section 6.6 gives that $X_c = (0.7 * \frac{18+2*9+3*6+4*5}{38}) - 0.2 \approx 1.16$ for the shelter and $X_c = (0.7 * \frac{20+2*3+3*5+4*1}{29}) - 0.2 \approx 0.89$ for the regular class. Since both values of the X_c are smaller than the half of its value in the ideal case (3.3: 2 = 1.65) the two classes demonstrated again a less than satisfactory quality performance, with the performance of the shelter class being better.

It was logically expected that the test's results for the shelter class would be worse than in the regular class. However, the situation was opposite. Surprisingly, the shelter class performed better. This happened because the teacher, taking into account that the students in this class were not proficient in English, worked on a daily basis on developing the students' mathematics vocabulary and comprehension in reading mathematics content problems. This training affected student's CrT and PS abilities.

Test's Questions

1. Sketch a graph to model each of the following four situations. Think about the shape of the graph and whether it should be a continuous line or not.

A: Candle

Each hour a candle burns down the same amount. x = the number of hours that have elapsed. y = the height of the candle in inches.

B: Letter

When sending a letter, you pay quite a lot for letters, weighing up to an ounce. You then pay a smaller, fixed amount for each additional ounce (or part of an ounce.)

x = the weight of the letter in ounces.

y = the cost of sending the letter in cents.

C: Bus

A group of people rent a bus for a day. The total cost of the bus is shared equally among the passengers.

x = the number of passengers.

y = the cost for each passenger in dollars.

D: Car value

My car loses about half of its value each year.

x = the time that has elapsed in years.

y = the value of my car in dollars.

2. The formulas below are models for the situations. Which situation goes with each formula? Write the correct letter (A, B, C or D) under each one.

$y = \frac{300}{x}$, $y = 12 - 0.5x$. $y = 30 + 20x$. $y = 2000 \cdot (0.5)^x$.

3. Answer the following questions using the formulas. Under each answer show your reasoning.

How long will the candle last before it burns completely away?
How much will it cost to send a letter weighing 8 ounces?

If 20 people go on the coach trip, how much will each have to pay?

How much will my car be worth after 2 years?

EVALUATING STUDENT UNDERSTANDING OF THE INFINITY

Conceptions of the Infinity

Philosophers, mathematicians, mathematical historians and educators, students and many others have struggled for centuries to resolve the various issues and paradoxes regarding conceptions of the infinity. Aristotle's (384-322 BC) ***potential/actual dichotomy of the infinity*** dominated these conceptions for centuries. According to Aristotle the potential infinity could be understood as the infinite presented over time, while the actual infinity is the infinite present at a moment in time. The actual infinity is incomprehensible, because the underlying process of such an actuality would require the whole of time. This distinction of the concept of infinity allowed Aristotle to acknowledge the existence of the infinite, provided that it was not present "all at once" (Moore, 1999: p. 39). Further, the actual infinity explains all the paradoxes connected to the infinite.

However, views also appeared disputing the ideas of Aristotle, mainly expressed by the ***rationalists***, who believed that we can invoke the pure logic for the understanding of the real world in general and the actual infinity in particular. Bolzano (1741-1848) advanced, against the empiricist Aristotle's negative assertion, the idea of the existence of an ***infinite collection*** as a completed whole. His main argument to support this view was the existence of the ***large finite numbers***, like the grains of sand in a desert, a set with 10 $10^{10^{10^{10}}}$ elements, etc, which, although they doubtlessly exist, they cannot been enumerated by human beings. However, one concern with Bolzano's view is that the examples he used are finite sets. For instance, in case of enumerating the set of the first 10 $10^{10^{10^{10}}}$ natural numbers one can reflect on the last counting number as indicating its cardinality, a fact which cannot occur in an infinite set, where there is no such number.

Cantor (1845-1918) extended Bolzano's thinking. His theory of ***transfinite numbers*** is connected to his view that infinite sets to which a cardinality or order can be assigned "enjoy a kind of finitude" or are "really finite". Cantor thus suggests three cognitive categories, the ***finite***, the ***attainably infinite*** and the ***unattainably infinite***. The last one, termed by Moore (1999) as the "really infinite", refers to immeasurably large collections to which no cardinality or order can be assigned, like the collection of everything thinkable, the set of all the sets, etc. According to Cantor, actual infinite entities are considered to be attainably infinite, while potentially infinite collections that cannot be actualized are considered to be unattainably infinite.

Nowadays, the best way for connecting the potential to the actual infinity is probably the use of ***fractals*** (Mandelbrot, 1983), which are obtained by infinite processes characterized by a kind of self – similarity. Consider, for example, the ***ternary set*** discovered by Henry John Stephen Smith in 1874, but better known as *the* "Cantor's comb or dust". This set, through the consideration of which Cantor (1883) and others helped for laying the foundations of the modern point-set Topology, is created by removing repeatedly the open middle thirds of a line segment (Wikipedia, 2015). The first five steps of this construction are represented in Figure 7.

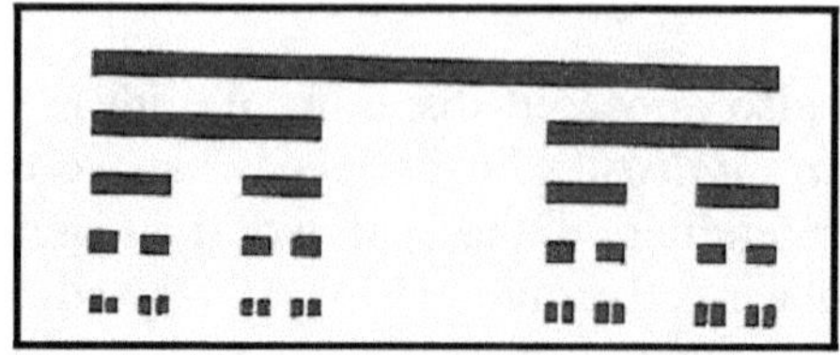

Figure 7: Graph of the ternary set

Figure 7 does not represent the set's final image, the creation of which requires an infinite number of such steps (actual infinity); it gives however a very precise approximation of it. In fact, it is easy to observe that the left and right parts of Figure 20 are similar, containing equal lengths. Further, each of these parts is similar to the whole figure and it also contains its own left and right parts. Therefore we have 4, 8, 16, ,,,,,, etc., smaller subsets similar to the original set. As the process continues, it becomes evident that the ternary set contains an infinite number of smaller and smaller subsets, all of which are similar to the original set (self-similarity).

Cantor's comb is probably the first fractal discovered in the history of mathematics.

However, although nowadays in just about every case there is a rigorous mathematical explanation, many students have considerable difficulty in understanding the infinite. Tsamir (2001), for example, found that prospective teachers erroneously attribute properties of finite to infinite sets, Mamona-Downs (2001) found that many students consider that the limit of a sequence is its last term and, given the sequence (a_n), $n \in \mathbb{N}$, they write a_∞ for its limit, etc.

The Classroom Experiment

One can find in the literature reflections of the development of the concept of infinity in students of today (Tsamir, 2001, Hauchart & Rouche, 1987, Nunez, 1993, etc). The pioneer of this study was E. Fischbein, whose empirical researches revealed many conflicting intuitional student perceptions of the infinite (.Fischbein, 1978, 1987, 2001, Fischbein, Tirosh

& Melamed, 1981, 1983). His last article (Fischbein, 2001) was published just after his death, in 2001, together with six articles of other authors (Tsamir, 2001, Mamona-Dawns, 2001, Jahnke, 2001, Kleiner, 2001, Monagham, 2001, Tall, 2001) in a special issue of the "Educational Studies of Mathematics" on the concept of infinity, dedicated to his memory.

The impulsion to perform the following classroom experiment was given by our concern to study the effects that an instructor's lecture to students on the basic philosophical/epistemological aspects of the infinity could have for the improvement of their abilities to deal successfully in their mathematical courses with situations involving the concept of infinity. For this, we selected two equivalent student groups - according to the marks obtained in their first term course "Higher Mathematics I"- from the School of Technological Applications (prospective engineers) of the Graduate Technological Educational Institute (T. E. I.) of Western Greece being at their second term of studies. A two hours lecture was delivered separately to the students of each group. The lecture for the first (experimental) group was focused mainly on the basic philosophical /epistemological aspects concerning the infinity (Section 6.11.1). On the contrary, the attention of the lecture for the second (control) group was turned to more examples to the topics of the course "Higher Mathematics I" involving, directly or indirectly, the concept of infinity. Note that the course involves an introductory chapter on the basic sets of numbers,

Differential and Integral Calculus in one variable, elements of Analytic Geometry and Linear Algebra.

Next, a written test was performed for both groups in terms of the questionnaire that will be presented below together with some representative wrong student answers. The student answers were marked in a climax from 0 to 100 and the scores obtained were the following:

Experimental group **(G_1):** 100(2 times), 99(1), 98(2), 95(3), 94(2), 92 (3), 90(2), 89(1), 88(3), 85(1), 82(2), 80(4), 78(3), 76(2), 75(4), 72(3), 70(1), 68(2), 60(1), 58(2), 57(1), 56(2), 55(2), 54(1), 50(2), 45(3), 42(2), 40(2), 35(1).

Control group **(G_2)*:*** 100(2), 99(1), 98(1), 97(1), 95(2), 92(4), 91(1), 90(2), 88(1), 85(5), 82(2), 80(6), 78(9), 75(13), 70(3), 64(4), 60(8), 58(2), 56(3), 55(3), 50(7), 45(2), 40(3). The above data are depicted in Table 7:

Table 7: Characterization of the student performance

Characterizations	G_1	G_2
A	20	20
B	15	30
C	7	15
D	10	15
F	8	5
Total	60	85

The evaluation of the above data was performed in the following ways:

i) Mean values: A straightforward calculation gives that the means of the student scores are approximately 73.28 and 71.91 for the experimental and the control group respectively. This shows that the mean performance of both groups was good (C), with the performance of the experimental group being better.

ii) GPA index: Replacing the data of Table 12 to formula (1) of Section 6.1 and making the corresponding calculations one finds that the GPA

index is 2.48 for the experimental and 2.52 for the control group. These values, being greater than half of the GPA's maximal value (4:2 = 2) indicate a more than satisfactory quality performance for both groups. Nevertheless, the control group demonstrated a slightly better performance.

iii) RFAM: Applying the first of formulas (4) of Section 6.3 on the data of Table 12 one finds that $x_c \approx 2.98$ for the experimental and $x_c \approx 3.02$ for the control group respectively. Therefore, according to the criterion of Section 6.3 the control group demonstrated a better performance. Further, since the value of x_c for both groups was greater than half of its value in the ideal case ($\frac{9}{2}$: 2 = 2.25), their quality performance is characterized as more than satisfactory.

iv) GRFAM (or TFAM or TpFAM): The first of formulas (7) of Section 6.6 gives that $X_c \approx 2.22$ for the experimental and $X_c \approx 2.27$ for the control group. Since both values of the X_c are greater than half of its value in the ideal case (3.3: 2 = 1.65) the two groups demonstrated a more than satisfactory quality performance, with the performance of the control group being better.

In concluding, it was normally expected for the control group to demonstrate a better performance than the experimental one, since its students were exposed during the two-hour extra lecture to more examples concerning the infinity. However, according to the mean values of the student scores the experimental group demonstrated a better mean performance than the control group. On the contrary, according to the values of the GPA index, and the outcomes of RFAM and of its equivalent variations GRFAM, TFAM and TpFAM, the control group demonstrated a slightly better quality performance. Therefore, the instructor's extra lecture to the students of the experimental group on the basic philosophical/epistemological aspects of the infinity enhanced the performance of the mediocre students (lower scores), but it had a much smaller effect on the performance of the good students (higher scores), who had already (before the additional two-hour lecture) understood well the concept of the infinity.

Questionnaire of the Experiment

1. *a) Compare the numbers 4.9999..... and 5.*

b) Are there any fractions between $\frac{1}{10}$ *and* $\frac{1}{11}$*? If yes, write one of them.*

2. Compare the cardinalities of the sets N of natural numbers, N_E *of the even natural numbers, Z of the integers, Q of the rational and R of the real numbers. Justify your answers.*

3. Examine if there exist the limits: a) $\lim_{x\to 2}\sqrt{x^2-9}$, *b)* $\lim_{x\to a} f(x)$, *with* $f(x)=\begin{cases}1, & x\in Q\\ 0, & x\in R-Q\end{cases}$, $a\in R$, *where Q is the set of rational and R is the set of real numbers.*

4. Given the line segment AB with length 1 m we add to it the line segments BC of length $\frac{1}{2}$ *m, CD of length* $\frac{1}{4}$ *m, DE of length* $\frac{1}{8}$ *m, EG of length* $\frac{1}{16}$ *m,.... and so on. Find the total length of AB + BC + CD + DE + EG +.....* (This problem was retrieved from Fischbein, 1978).

5. Starting from the interval [0, 1] we delete first its middle third ($\frac{1}{3}$, $\frac{2}{3}$*), then the middle thirds (*$\frac{1}{9}$,$\frac{2}{9}$*) and (*$\frac{7}{9}$,$\frac{8}{9}$*) of the two remaining intervals [0,* $\frac{1}{3}$*] and [*$\frac{2}{3}$*, 1] respectively, and so on (ternary set).*

a) Find the total length of the removed intervals when the above process is repeated infinitely many times (the lengths of the removed intervals form a decreasing geometric progression with first term equal to $\frac{1}{3}$ and ratio $\frac{2}{3}$, therefore the sum of all these lengths is equal to $\frac{\frac{1}{3}}{1-\frac{2}{3}}$= 1).

b) Are there any points left behind in this case?

Wrong Student Answers

1. a) 5 is greater than 4.9999……

b) No, because $\frac{1}{11}$ is the next fraction following $\frac{1}{10}$.

2. Since $N_E \subseteq N$, N has a greater cardinality, etc. Also: All these sets are infinite and therefore they have the same cardinality, which is equal to ∞, or they have no cardinality, which, in case of existence, should be a real number.

3. a) The limit does not exist, because $2^2 - 9 < 0$ and the negative numbers have not real square roots.

b) There are two limits equal to 0 and 1 respectively.

4. The total length is infinite, since the successive additions are repeated infinitely many times.

5. a) The total length removed is less than 1, because there are some points of the initial interval [0, 1] left behind, like $\frac{1}{3}$, $\frac{2}{3}$, etc.

b) There are no points left behind, since the total length removed is equal to 1 (these students answered correctly part (a) of this question).

OTHER APPLICATIONS

Other applications of the GOC technique as an assessment method (RFAM) involve the Zone of Proximate Reasoning (Subbotin et al., 2011), CBR (Subbotin & Voskoglou, 2011, Voskoglou, 2013a), AR (Voskoglou & Subbotin, 2012), , PS (Voskoglou, 2012b), MM (Voskoglou, 2013b) Computational Thinking (CT) (Voskoglou, 2015a), DM (Voskoglou, 2014a), the APOS/ACE instructional treatment of Mathematics (Voskoglou, 2015b), the Bloom's Taxonomy (Voskoglou & Subbotin, 2015a), etc.

Other applications of the RFAM's equivalent variations GRFAM, TFAM and TpFAM to assessment processes presented in earlier works involve the process of learning (Subbotin, 2014, Subbotin & Bilotskii, 2014), PS (Voskoglou, 2014c), Faculty evaluations by students (Subbotin & Voskoglou, 2015), AR (Voskoglou & Subbotin, 2015b), CBR (Voskoglou & Subbotin, 2015c), the Bloom's taxonomy (Voskoglou & Subbotin, 2015), Bridge players' performance (Voskoglou, 2015c), etc

CONCLUSIONS

The use of the COG defuzzification technique as an assessment method (RFAM) was developed in this Chapter and its applications to a variety of human activities were described. In contrast to the measurement of a system's uncertainty, which focuses on the assessment of its mean performance, RFAM focuses on its quality performance by assigning greater coefficients to the higher scores.

The equivalent to each other GRFAM, TFAM and TPFAM were also developed as variations of the RFAM and they were applied for assessing CrT skills and the student difficulties for understanding the concept of infinity. However, it was finally proved that the above models provide always the same assessment conclusions with RFAM on comparing the performance of two different groups. Further it was proved that, if the values of GPA index for two groups are different, then the above methods provide the same assessment conclusions with the GPA index as well. On the contrary, if the values of the GPA index are the same, then the conclusions obtained by it could not be logically based. In such situations RFAM and its variations, due to their logical nature, become useful.

REFERENCES

Borowski, E. J. & Borwein, J. M. (1991), *The Harper Collins Dictionary of Mathematics*, Harper Resource, New York.

Bruner, J. (1960), *The Process of Education*, University Press, Cambridge, MA, Harvard.
Dictionary.reference.com (2013), Critical thinking, retrieved on February 12, 2013 from: http://dictionary.reference.com/browse/criticalthinking

Fischbein, E. (1978), Intuition and mathematical education, *Osnabrucker Schriften zur Mathematik*, 1, 148-176.

Fischbein, E. , Tirosh, D. & Hess, P. (1979), The intuition of infinity, E*ducational Studies in Mathematics*, 10, 3-40.

Fischbein, E. , Tirosh, D. & Melamed, U. (1981), Is it possible to measure the intuitive acceptance of mathematical statement?, E*ducational Studies in Mathematics*, 12, 491-512.

Fischbein, E. (1987), *Intuition in Science and Mathematics,* Reidel Publishing, Kluwer Academic Publishers, Dordrecht, Netherlands.

Fischbein, E. (2001), Tacit Models and Infinity, E*ducational Studies in Mathematics*, 48, 309-329.

Greiner, C., Serdyukov, P., Subbotin, I., & Serduykova, N. (2004), Enhancing E-learning outcomes through iteration, *E-Learn World Conference on E-Learning in Corporate, Government, Healthcare, & Higher Education*, Washington, D.C., November.

Halpern, D. & Hakel (2004), M. Applying the science of learning to the university and beyond: Teaching for long-term retention and transfer, Presentation at *80th WASC Annual Meeting*, San Jose, April, 14-16.

Hauchart, C. & Rouche, N. (1987), *Apprivoiser l'infini: Un enseignement des debuts de l'analyse,* CIACO, Louvain.

Jahnke, H. N. (2001), Cantor's Cardinal and Ordinal Infinities: An Epistemological and Didactic View, E*ducational Studies in Mathematics*, 48, 175-197.

Kaplan, E. (1964), *Winning Contract Bridge Complete,* Fleet Publishing Corporation, N. Y.

Kleiner, I. (2001), History of the Infinitely Small and the Infinitely Large in Calculus, E*ducational Studies in Mathematics*, 48, 137-174.

Mamona-Downs, I. (2001), "Letting the Intuitive Bear on the Formal: A Didactical Approach for the Understanding of the Limit of a Sequence", E*ducational Studies in Mathematics*, 48, 259-288.

Mandelbrot, Benoit, B. (1983), *The Fractal Geometry of Nature*, W. H. Freeman and Company.

McPeck, J. (1992), Thoughts on subject specificity. In S. Norris (Ed.), *The generalizability of critical thinking* (pp. 198–205), New York: Teachers College Press.

Monagham, J. (2001), Young Peoples' Ideas of Infinity, E*ducational Studies in Mathematics*, 48, 239-257.

Moore, A. W. (1999), *The Infinite*, 2nd ed., Routledge and Paul, London

Nunez Errazuriz, R. (1993), *En deca detranfini,* Editions Universitaires, Fribourg.

Opperman, R. & Thomas, C. (2004), Learning and Problem Solving as an Iterative Process: Learners' Living Repository: LEAR. http://ui4all.ics.forth.gr/UI4ALL-95/oppermann.pdf, retrieved April, 4.

Pagat.com (2014), Bridge rules and variations, retrieved on February 2014 from www.pagat.com/boston/bridge.html

Paul, R. (1982), Teaching critical thinking in the strong sense: A focus on self-deception, world views and a dialectical mode of analysis, I*nformal Logic Newsletter*, 4(2), 2-7.

Rifkin, J. (2016), The Third Industrial Revolution, retrieved on April 18, 2016 from www.thethirdindustrialrevolution.com .

Subbotin, I., Badkoobehi, H. & Bilotskii, N. (2004), Application of Fuzzy Logic to Learning Assessment, *Didactics of Mathematics: Problems and Investigations*, 22, 38-41.

Subbotin, I. Ya., Mossavar-Rahmani, F. & Bilotskii, N. N. (2006), Fuzzy logic and iterative assessment, *Didactics of Mathematics: Problems and Investigations,* 25, 221-227.

Subbotin, I., Mossavar-Rahmani, F. & Bilotskii, N. (2011), Fuzzy logic and the concept of the Zone of Proximate Development, *Didactics of Mathematics: Problems and Investigations*, 36, 101-108.

Subbotin, I.Ya. & Voskoglou, M.Gr. (2011), Applications of Fuzzy Logic to Case-Based Reasoning, *International Journal of Applications of Fuzzy Sets and Artificial Intelligence,* 1, 7-18.

Subbotin, I.Ya. (2014), Trapezoidal Fuzzy Logic Model for Learning Assessment, *arXiv 1407.0823[math.GM]*.

Subbotin, I.Ya. & Voskoglou, M.Gr. (2014a), Language, Mathematics and Critical Thinking: The Cross Influence and Cross Enrichment, *Didactics of Mathematics: Problems and Investigations,*41, 89-94.

Subbotin, I.Ya. & Voskoglou, M.Gr. (2014b), A Triangular Fuzzy Model for Assessing Critical Thinking Skills, *International Journal of Applications of Fuzzy Sets and Artificial Intelligence*, 4, 173 -186.

Subbotin, I.Ya. & Voskoglou, M.Gr. (2014c), Fuzzy Assessment Methods, *Universal Journal of Applied Mathematics,* 2(9), 305-314.

Subbotin, I.Ya. & Bilotckii, N. N. (2014), Triangular fuzzy logic model for learning assessment, *Didactics of Mathematics: Problems and Investigations*, 41, 84-88.

Subbotin, I.Ya. (2015), On Generalized Rectangular Fuzzy Model for Assessment, *Global Journal of Mathematics*, 2(1), 65- 70.

Subbotin, I.Ya. & Voskoglou, M.Gr. (2015), Fuzzy models for Assessing Faculty Evaluations, *International Journal of Applications of Fuzzy Sets and Artificial Intelligence,* 5, 5-22.

Subbotin, I.Ya. & Voskoglou, M.Gr. (2016), An Application of the Generalized Rectangular Fuzzy Model to Critical Thinking Assessment, *American Journal of Educational Research*, 4(5), 397- 403.

Swinburne.edu.au (2014), Grade Point Average Assessment, retrieved from http://www.swinburne.edu.au/studentadministration/assessment/gpa.html on October, 2014

Tall, D. (2001), Natural and Formal Infinities, *Educational Studies in Mathematics*, 48, 199-238.

Tsamir, P. (2001), "When the Same is not Perceived as Such: The Case of Infinite Sets", *Educational Studies in Mathematics*, 48, 289-307.

van Broekhoven, E. & De Baets, B. (2006), Fast and accurate centre of gravity defuzzification of fuzzy system outputs defined on trapezoidal fuzzy partitions, *Fuzzy Sets and Systems*, 157(7), 904-918.

Voskoglou, M. Gr. (1999), The process of learning mathematics: A fuzzy set approach, *Heuristic and Didactics of Exact Sciences*, 10, 9-13.

Voskoglou, M. Gr. (2012a), A study on fuzzy systems, *American Journal of Computational and Applied Mathematics*, 2(5), 232-240.

Voskoglou, M. Gr. (2012b), A Fuzzy Model for Problem Solving, *Turkish Journal of Fuzzy Systems,* 3(1), 1-15

Voskoglou, M.Gr. & Subbotin, I.Ya. (2012), Fuzzy Models for Analogical Reasoning, *International Journal of Applications of Fuzzy Sets and Artificial Intelligence,* 2, 19-38.

Voskoglou, M. Gr. (2013a), Case-Based Reasoning in Computers and Human Cognition: A Mathematical Framework, *International Journal of Machine Intelligence and Sensory Signal Processing,* 1, 3-22

Voskoglou, M. Gr. (2013b), Application of Fuzzy Logic to Systems' Modelling, *International Journal of Fuzzy Systems Applications,* 3(2), 1-15.

Voskoglou, M. Gr. (2014a), Probability and Fuzziness in Decision Making, *Egyptian Computer Science Journal,* 38(3), 86-99.

Voskoglou, M. Gr. (2014b), Assessing the Players' Performance in the Game of Bridge, *American Journal of Applied Mathematics and Statistics*, 2(3), 115-120.

Voskoglou, M. Gr. (2014c), A Triangular Fuzzy Model for Assessing Problem Solving Skills, *Annals of Pure and Applied Mathematics*, 7(1), 53-58.

Voskoglou, M. Gr. (2015a), An Application of Fuzzy Sets for Studying the Influence of Computational Thinking in Learning Mathematics, *Journal of Mathematical Sciences and Mathematics Education*, 10(1), 30-47.

Voskoglou, M. Gr. (2015b), Fuzzy Logic in the APOS/ACE Instructional Treatment of Mathematics, *American Journal of Educational Research*, 3(3), 330-339.

Voskoglou, M. Gr. (2015c), Variations of the COG Defuzzification Technique for Assessment Purposes, *Journal of Mathematical Sciences and Mathematics Education*, 10(2), 38-57.

Voskoglou, M.Gr. & Subbotin, I.Ya. (2015a), A Fuzzy Model for Measuring Student Learning Based on Bloom's Taxonomy, *Egyptian Computer Science Journal*, 39(2), 43-55.

Voskoglou, M.Gr. & Subbotin, I.Ya. (2015b), Application of the Triangular Fuzzy Assessment Model to Assessment of Analogical Reasoning Skills, *American Journal of Applied Mathematics and Statistics*, 3(1), 1-6.

Voskoglou, M.Gr. & Subbotin, I.Ya. (2015c), Evaluating the Effectiveness of a CBR System, *American Journal of Computational and Applied Mathematics*, 5(2), 27-32.

Wikipedia.org (2014a), Center of mass: Definition, retrieved on October, 2014 from http://en.wikipedia.org/wiki/Center_of_mass#Definition .

Wikipedia.org (2014b), Trapezoid: Other properties. Retrieved on October, 2014 from http://en.wikipedia.org/wiki/trapezoid#other_properties

Wikipedia.org (2014c), Center of mass: A system of particles, retrieved from http://en.wikipedia.org/wiki/Center_of_mass#A_system_of_particles , on October 10, 2014.

Wikipedia.org (2015), Cantor set, retrieved on December 21, 2015 from: https://en.wikipedia.org/wiki/Cantor_set

CHAPTER 7
Use of Fuzzy Numbers as Assessment Tools

ABSTRACT

Fuzzy Numbers play an important role in fuzzy mathematics analogous to the role played by the ordinary numbers in crisp mathematics. A method is developed in this Chapter of using the Triangular and the Trapezoidal Fuzzy Numbers as assessment tools, which is validated by the parallel use of the calculation of the mean values and of the GPA index. The examples presented concern assessment of student and basket – ball player performance. The background needed for our purposes from the theory of Fuzzy Numbers is exposed in the first part of the Chapter.

BASIC CONCEPTS

1. Definition: A ***Fuzzy Number*** (FN) is a FS A on the set ***R*** of real numbers with membership function m_A: ***R*** $\rightarrow$ [0, 1], such that:

- A is ***normal***, i.e. there exists x in ***R*** such that $m_A(x) = 1$,
- A is ***convex***, i.e. all its ***a-cuts*** $A^a = \{x \in U: m_A(x) \geq a\}$, a in [0, 1], are closed real intervals;
- Its membership function $y = m_A(x)$ is a piecewise continuous function.

2. Counter-example: The graph of a FS on ***R*** is presented in Figure 1, which is not convex and therefore it is not a FN. For example, it can be observed that $A^{0.4} = [5, 8.5] \cup [11, 13]$, therefore $A^{0.4}$ is not a closed interval.

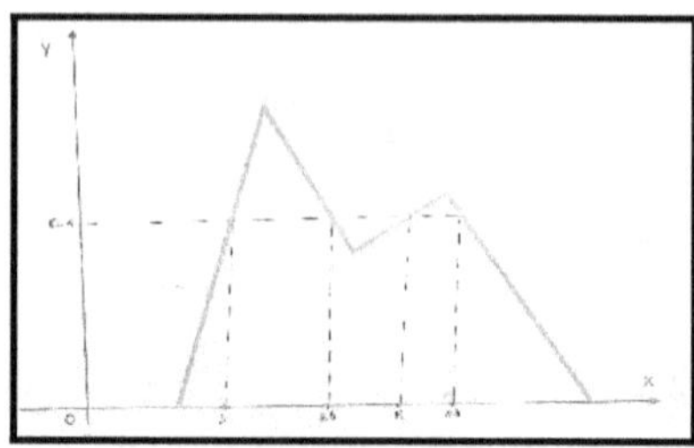

Figure 1: Example of a non convex fuzzy set on ***R***

Since the x-cuts A^x of a FN, say A, are closed real intervals, we can write $A^x=[A_l^x, A_r^x]$ for each x in [0, 1], where A_l^x, A_r^x are real numbers depending on x.

The following statement defines a *partial order* on the set of all FNs:

3. Definition: Given the FNs A and B we write $A \leq B$ (or $\geq$) if, and only if, $A_l^x \leq B_l^x$ and $A_r^x \leq B_r^x$ (or $\geq$) for all x in [0, 1]. Two such FNs are called ***comparable***, otherwise they are called ***non comparable.***

4. Remark: One can define the four basic ***arithmetic operations*** on FNS in the following two, equivalent to each other, ways (Kaufmann and Gupta, 1991):

i. With the help of their *a*-cuts and the Representation-Decomposition Theorem of Ralesscou - Negoita (Sakawa, 1993, Theorem 2.1, p.16) for FS. In this way the fuzzy arithmetic is turned to the well known arithmetic of the closed real intervals.

ii. By applying the Zadeh's extension principle (Klir & Folger, 1988, Section 1.4, p.20), which provides the means for any function f mapping a crisp set X to a crisp set Y to be generalized so that to map fuzzy subsets of X to fuzzy subsets of Y.

In practice the above two general methods of the fuzzy arithmetic, requiring laborious calculations, are rarely used in applications, where the utilization of simpler forms of FNs is preferred.

For general facts on FNs we refer to the classical on the subject book of Kaufmann and Gupta (1991).

TRIANGULAR FUZZY NUMBERS (TFNs)

A TFN (*a, b, c*), with *a, b, c* in ***R*** represents the fuzzy statement "the value of *b* lies in the interval [*a, c*]". The membership function of (*a, b, c*) is zero outside the interval [*a, c*], while its graph in [*a, c*] consists of two straight line segments forming a triangle with the OX axis (Figure 2).

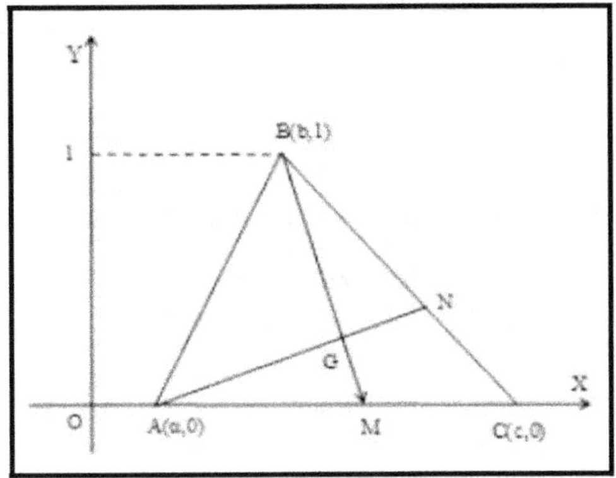

Figure 2: Graph and COG of the TFN (a, b, c)

Therefore the analytic definition of a TFN is given as follows:

5. Definition: Let *a, b* and *c* be real numbers with $a < b < c$. Then the TFN
(*a, b, c*) is a FN with membership function:

$$y = m(x) = \begin{cases} \frac{x-a}{b-a}, & x \in [a,b] \\ \frac{c-x}{c-b}, & x \in [b,c] \\ 0, & x < a \text{ or } x > c \end{cases}$$

The following two Propositions refer to basic properties of TFNs that they are going to be used later in this work:

6, Proposition: The x-cuts A^x of the TFN A = (a, b, c), $x \in [0, 1]$, are calculated by the formula $A^x = [a + x(b - a), c - x(c - b)]$.

Proof: The x-cut $A^x = \{y \in \boldsymbol{R}: m(y) \geq x\}$ of the TFN A is a real interval of the form $A^x = [A_l^x, A_r^x]$, with A_l^x in [a, b] and A_r^x in [b, c] such that $m(A_l^x) = m(A_r^x) = x$. Therefore, by Definition 5 we have that $\frac{A_l^x - a}{b - a} = x \Leftrightarrow A_l^x = a + x(b-a)$.

Similarly, by Definition 5 again, $\frac{c - A_r^x}{c - b} = x \Leftrightarrow A_r^x = c - x(c - b)$.

7. Proposition (Defuzzification of a TFN): The coordinates (X, Y) of the COG of the graph of the TFN (a, b, c) are calculated by the formulas $X = \frac{a+b+c}{3}$,
$Y = \frac{1}{3}$.

Proof: The graph of the TFN (a, b, c) is the triangle ABC of Figure 16, with A$(a, 0)$, B$(b, 1)$ and C $(c, 0)$. Then, the COG, say G, of ABC is the intersection point of its medians AN and BM. The proof of the Proposition is easily obtained by calculating the equations of AN and BM and by solving the linear system of these two equations.

8. Arithmetic Operations on TFNs: It can be shown (Kaufmann and Gupta, 1991) that the two general methods of defining arithmetic operations on FNs mentioned in Remark 4 lead to the following simple rules for the ***addition*** and ***subtraction*** of TFNs:

Let A = (a, b, c) and B = (a_1, b_1, c_1) be two TFNs. Then

- The sum A + B = $(a+a_1, b+b_1, c+c_1)$.
- The difference A - B = A + (-B) = $(a-c_1, b-b_1, c-a_1)$, where –B = $(-c_1, -b_1, -a_1)$ is defined to be the *opposite* of B.

In other words, the opposite of a TFN, as well as the sum and the difference of two TFNs are always TFNs. On the contrary, the ***product*** and the ***quotient*** of two TFNs, although they are FNs, they are not always TFNs, unless if a, b, c, a_1, b_1, c_1 are in $\boldsymbol{R}^+$ (Kaufmann and Gupta, 1991).

The following two ***scalar operations*** can be also defined:

- k + A= (k+a, k+b, k+c), k$\in$$\boldsymbol{R}$
- kA = (ka, kb, kc), if k>0 and kA = (kc, kb, ka), if k<0.

TRAPEZOIDAL FUZZY NUMBERS (TPFNs)

A TpFN (a, b, c, d) with a, b, c, d in $\boldsymbol{R}$ represents the fuzzy statement "approximately in the interval [b, c]". Its membership function $y=m(x)$ is zero outside the interval [a, d], while its graph in the interval [a, d] is the union of three straight line segments forming a trapezoid with the X-axis (see Figure 3),

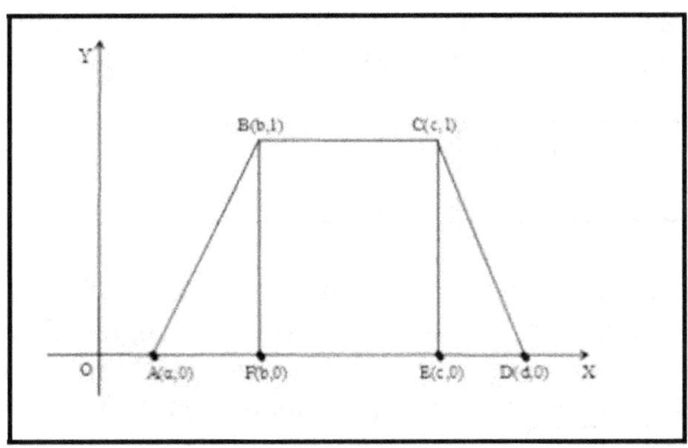

Figure 3: Graph of the TpFN (a, b, c, d)

Therefore, the analytic definition of a TpFN is given as follows:

9. Definition: Let $a < b < c < d$ be given real numbers. Then the TpFN
(a, b, c, d) is the FN with membership function:

$$y = m(x) = \begin{cases} \frac{x-a}{b-a}, & x \in [a,b] \\ x = 1, & x \in [b,c] \\ \frac{d-x}{d-c}, & x \in [c,d] \\ 0, & x < a \text{ and } x > d \end{cases}$$

It is easy to observe that the TpFNs are generalizations of TFNs. In fact, the TFN (*a, b, d*) can be considered as a special case of the TpFN (*a, b, c, d)* with *b=c.*

***10. Arithmetic Operations on TpFNs*:** It can be shown (Kaufmann and Gupta, 1991) that the addition and subtraction of two TpFNs are performed in the same way as for TFNs (paragraph 8). Also, the two scalar operations defined in paragraph 8 for TFNs, hold also for TpFNs.

***11. Proposition (Defuzzification of a TpFN)**:* The coordinates (*X, Y*) of the COG of the graph of the TpFN (*a, b, c, d*) are calculated by the formulas

$$X = \frac{c^2 + d^2 - a^2 - b^2 + dc - ba}{3(c+d-a-b)}, \quad Y = \frac{2c+d-a-2b}{3(c+d-a-b)}.$$

Proof: We divide the trapezoid forming the graph of the TpFN (*a, b, c, d*) in three parts, two triangles and one rectangle (Figure 17). The coordinates of the three vertices of the triangle ABE are (*a,* 0), (*b,* 1) and (*b,* 0) respectively, therefore by Proposition 7 the COG of this triangle is the point C_1 ($\frac{a+2b}{3}, \frac{1}{3}$). Similarly one finds that the COG of the triangle FCD is the point C_2 ($\frac{d+2c}{3}, \frac{1}{3}$). Also, it is easy to check that the COG of the rectangle BCFE, being the intersection point of its diagonals, is the point C_3 ($\frac{b+c}{2}, \frac{1}{2}$). Further, the areas of the two triangles are equal to

$S_1 = \frac{b-a}{2}$ and $S_2 = \frac{d-c}{2}$ respectively, while the area of the rectangle is equal to S_3 = c- b.

The coordinates of the COG of the trapezoid, being the resultant of the COGs C_i (x_i, y_i), for i=1, 2, 3, are calculated by the formulas $X = \frac{1}{S}\sum_{i=1}^{3} S_i x_i$, $Y = \frac{1}{S}\sum_{i=1}^{3} S_i y_i$ (1), where $S = S_1 + S_2 + S_3 = \frac{c+d-b-a}{2}$ is the area of the trapezoid (Wikipedia, 2014c).

The proof of the Proposition is completed by replacing the above found values of S, S_i, x_i and y_i, i = 1, 2, 3, in formulas (1) and by performing the corresponding operations.

The following definition is introduced to enable the use of the TFNs and the TpFNs as assessment tools:

12. Definition: Let $A_i = (a_{1i}, a_{2i}, a_{3i}, a_{4i})$, i = 1, 2,…, n be TFNs (TpFNs), where n is a non negative integer, $n \geq 2$. Then we define the ***mean value*** of the A_i's to be the TFN (TpFN) $A = \frac{1}{n}(A_1 + A_2 + \ldots + A_n)$.

ASSESSING THE RESULTS OF THE APOS/ACE INSTRUCTIONAL TREATMENT FOR TEACHING/LEARNING MATHEMATICS

The APOS/ACE Theory

The ***APOS/ACE*** instructional treatment for learning and teaching mathematics was developed in the USA during the 1990's by a team of mathematicians and mathematics educators led by Ed Dubinsky (Asiala et al., 2001, Dubinsky & McDonald, 2001) , etc].

APOS is a theory based on Piaget's principle that an individual learns by applying certain mental mechanisms to build specific mental structures and uses these structures to deal with problems connected to the corresponding situations (Piaget, 1970). Thus, according to the APOS analysis, an individual deals with a mathematical situation by using certain mental mechanisms to build cognitive structures that are applied to the situation. These mechanisms are called ***interiorization*** and ***encapsulation*** and the related cognitive structures are ***Actions, Processes, Objects*** and ***Schemas***. The first letters of the last four words constitute the acronym APOS.

The theory postulates that a mathematical concept begins to be formed as one applies transformations on certain entities to obtain other entities. A transformation is first conceived as an action. For example, if an individual can think of a function only through an explicit expression and can do little more than substitute for the variable in the expression and manipulate it, he/she is considered to have an action understanding of functions.

As an individual repeats and reflects on an action, this action may be interiorized to a mental process. A process performs the same operation as the action, but wholly in the mind of the individual enabling her/him to imagine performing the transformation without having to execute each step explicitly. For example, an individual with a process understanding of a function thinks about it in terms of inputs, possibly unspecified, and transformations of those inputs to produce outputs.

When one becomes aware of a mental process as a totality and can construct transformations acting on this totality, then we say that the individual has encapsulated the process into a cognitive object. In case of functions encapsulation allows one to form sets of functions, to define operations on such sets, to equip them with a topology, etc.

Although a process is transformed into an object by encapsulation, this is often neither easy nor immediate. This happens because encapsulation entails a radical shift in the nature of one's conceptualization, since it signifies the ability to think of the same concept as a mathematical entity to which new, higher-level transformations can be applied. On the other hand, the mental process that led to a mental object through encapsulation remains still available and many mathematical situations require one to ***de-encapsulate*** an object back to the process that led to it. This cycle may be repeated one or more times. For example, in defining the sum f + g of two functions possessing a common domain, say A, it is necessary to reconsider again f and g at a process level and thinking of all x in A to obtain a new process associating to each x in A the sum f (x) + g(x). Then this new process must be encapsulated, in order to obtain the function f + g at an object level.

A mathematical topic often involves many actions, processes and objects that need to be organized into a coherent framework that enables the individual to decide which mental processes to use in dealing with a mathematical situation. Such a framework is called a schema. In the case of functions, for example, it is the schema structure that is used to see a function in a given mathematical or real-world situation.

However, one must notice that there are not any rubrics in general to assess explicitly the level of understanding (by students) of mathematics corresponding to each cognitive level (structure) of the APOS theory. This is in fact a matter depending on the instructor's experience and intuition.

The implementation of the APOS as a framework for teaching and learning a mathematical topic requires a theoretical analysis of the concepts under study, called ***Genetic Decomposition (GD)***. A GD comprises a description that includes mental constructions (actions, processes and objects) and the order in which it may be best for learners to experience them (Dubinsky & McDonald, 2001). The APOS theory has important consequences for education. Simply put, it says that the teaching of mathematics should consist of helping students use the mental structures they already have to develop an understanding of as much mathematics as those available structures can handle. For students to move further, teaching should help them to build new, more powerful structures for handling more and more advanced mathematics.

Dubinsky and his collaborators realized that for each mental construction that comes out of an APOS analysis, one can find a computer task of writing a program or code, such that, if a student engages in that task, he (she) is fairly likely to build the mental construction that leads to learning the mathematics. In other words, performing the task is an experience that leads to one or more mental constructions. As a consequence of the above finding, the pedagogical approach based on APOS analysis, known as the ***ACE teaching cycle***, is a repeated cycle of three components: ***Activities*** on the computer, ***Classroom*** discussion and ***Exercises*** done outside the class.

The target of the activities on the computer is to help students in building the proper mental constructions for the better understanding and learning of the corresponding mathematical topic. The students discuss later in the classroom their experiences from the computer tasks performed in the laboratory, they repeat the same tasks without the help of computer and they reach, under their instructor's guidance and help, to the proper conclusions. Finally, the purpose of the exercises, which are given by the tutor as a home work, is to check and to embed better the new mathematical knowledge [1, 2]. The implementation of the ACE cycle and its effectiveness in helping students making mental constructions and learn mathematics has been reported in several research studies of the Dubinsky's team (Weller et al., 2003, 2009, 2011, etc.) and by other researchers too (Maharaj, 2013, Voskoglou, 2013, etc.).

Assessing the APOS/ACE results

The following experiment took place recently in the city of Patras with subjects the students of two different Departments of the School of Management and Economics of the Graduate Technological Educational Institute (T. E. I.) of Western Greece attending the course "Mathematics for Economists I" (Calculus and Linear Algebra) of their first term of studies and having the same instructor. The students of both Departments had the same mathematical background from secondary education and the grades, which they had obtained in the mathematics exam for entrance in higher education, were of about the same level for the two Departments. Also, since they were in their first term of studies, they had attended no previous mathematical courses at the T. E. I. of Western Greece.

The teaching procedure involved four didactic hours (45 minutes each) per week for each Department. For the ***control*** Department (D_1) the lectures were performed in the traditional way on the board, followed by a number of exercises and problems with the students participating for their solutions. On the contrary, for the ***experimental*** Department (D_2) half of the teaching hours were spent in a computer laboratory and the rest in the classroom according to the motive of the APOS/ACE instruction. Since working with computers needs more time, the students of the control group were exposed to the solution of more exercises and problems.

At the end of the term the students of both groups participated in the same final exam and the scores obtained in a climax from 0 to 100 were the following:

Department D_1: 100(2 times), 99(3), 98(5), 95(8), 94(7), 93(1), 92 (6), 90(5), 89(3), 88(7), 85(13), 82(6), 80(14), 79(8), 78(6), 76(3), 75(3), 74(3), 73(1), 72(5), 70(4), 68(2), 63(2), 60(3), 59(5), 58(1), 57(2), 56(3), 55(4), 54(2), 53(1), 52(2), 51(2), 50(8), 48(7), 45(8), 42(1), 40(3), 35(1).

-

Department D_2: 100(1), 99(2), 98(3), 97(4), 95(9), 92(4), 91(2), 90(3), 88(6), 85(26), 82(18), 80(29), 78(11), 75(32), 70(17), 64(12), 60(16), 58(19), 56(3), 55(6), 50(17), 45(9), 40(6).

The student performance was characterized by the fuzzy linguistic labels (grades) A, B, C, D and F defined above. The two Departments performance was assessed by using the traditional methods of calculating the mean values and the GPA index first and by using TFNs next.

i) *Traditional methods:* It is straightforward to calculate the ***mean values*** of the student scores, which are equal to 73.79 and 72.39 for D_1 and D_2 respectively. Therefore, the two Departments demonstrated a good (C) ***mean performance***, with the performance of D_1 being better.

In order to calculate the values of the ***GPA index*** for the two Departments, Table 1 was formed by depicting the student performance in terms of the grades A, B, C, D and F defined above:

Table 1: Student performance in terms of the linguistic grades

Grade	D_1	D_2
A	60	60
B	40	90
C	20	45
D	30	45
F	20	15
Total	170	255

Replacing the data of Table 1 to formula (1) of Chapter 6 and making the corresponding calculations one finds the same value GPA = $\frac{43}{17} \approx$ 2.53 for the two Departments which indicates a more than satisfactory quality performance.

ii) Using TFNs: A TFN (denoted by the same letter) is assigned to each linguistic label (grade) as follows: A= (85, 92.5, 100), B = (75, 79.5, 84), C = (60, 67, 74), D= (50, 54.5, 59) and F = (0, 24.5, 49). The middle entry of each of the above TFNs is equal to the mean value of the student scores assigned to the corresponding grade. In this way a TFN corresponds to each student assessing his (her) individual performance. The representation of the linguistic labels A, B, C, D and F by TFNs has the advantage of determining numerically the scores corresponding to each label. In fact, the scores assigned to the above labels in the statement of this Example are not standard, since they may differ from case to case. In a more rigorous assessment, for instance, one could take A(90-100), B (80-89), C(70-79), D (60-69), F(<60), etc.

It becomes now clear that Table 1 gives rise to 170 TFNs representing the progress of the students of D_1 and 255 TFNs representing the progress of the students of D_2. Consequently, it is logical to accept that the overall performance of each Department is given by the corresponding mean value of the above TFNs (Definition 12). For simplifying our notation, let us denote the above mean values by the letter of the corresponding Department. Then, making straightforward calculations, one finds that

$$D_1 = \frac{1}{170}.(60A+40B+20C+30D+20F) \approx (63.53,\ 71.74,\ 79.94),$$

$$D_2 = \frac{1}{255}.(60A+90B+45C+45D+15F) \approx (65.88,\ 72.71,\ 79.53).$$

The left and right entries of the TFNs D_1 and D_2 show that the *mean performance* of the two Departments lies in the intervals (63.53, 79.74) and (65.88, 79.53) respectively, i.e. between being good (C) to very good (B).

Applying Proposition 6 one finds that the x-cuts of the above two TFNs are $D_1^x = [63.53+8.21x,\ 83.47-11.73x]$ and $D_2^x = [65.88+6.75x,\ 79.53-6.9x]$ respectively. But $63.53+8.21x \leq 65.88+6.75x \Leftrightarrow 1.46x \leq 2.35 \Leftrightarrow x \leq 1.61$, which is true, since x is in [0, 1]. On the contrary, $83.47-11.73x \leq 79.53-6.9x \Leftrightarrow 3.94 \leq 4.83x \Leftrightarrow 0.82 \leq x$, which is not true for all the values of x. Therefore, according to Definition 4, the TFNs D_1 and D_2 are not comparable, which means that in this way one can not decide which of the two Departments demonstrates the better performance.

To overcome this difficulty we defuzzify the TFNs D_1 and D_2. By Proposition 7, the COGs of the triangles forming the graphs of the TFNs

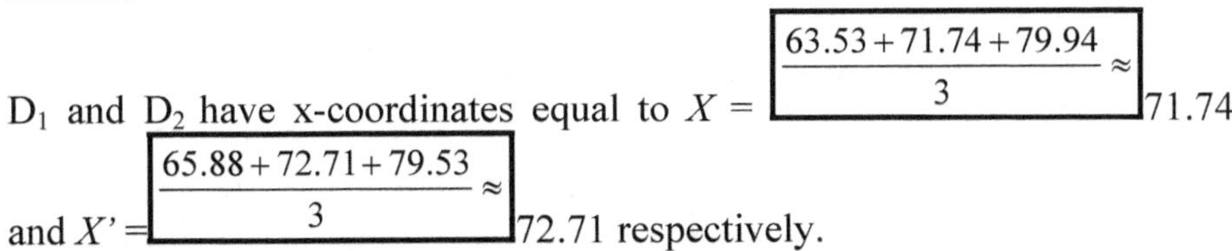

D_1 and D_2 have x-coordinates equal to $X = \frac{63.53+71.74+79.94}{3} \approx 71.74$ and $X' = \frac{65.88+72.71+79.53}{3} \approx 72.71$ respectively.

Observe now that the GOGs of the graphs of D_1 and D_2 lie in a rectangle with sides of length 100 units on the X-axis (student scores from 0 to 100) and one unit on the Y-axis (normal fuzzy sets). Therefore, *the nearer is the x-coordinate of the COG to 100, the better the corresponding Department's performance.* Thus, both Departments demonstrated a good mean performance, but, since $X < X'$, D_2 demonstrated a better mean performance than D_1.

Although both the calculation of the mean values and the use of TFNs measure the mean performance of the two Departments, the outcomes of the two methods lead to different conclusions concerning their performance. This is due to the different philosophy of the two methods, since the former is based on the principles of the traditional bi-valued logic, whereas the latter is based on the principles of the multi-valued FL.

The differences found between the performances of the two Departments are small enough to allow safe conclusions for possible advantages of the APOS/ACE teaching style with respect to the traditional, lecture-based method. However, the performance of the control Department was expected to be clearly better, since its students were exposed to the solution of more exercises and problems. Instead, the quality performance of the two Departments was found to be the same (GPA index), while the mean performance of the experimental Department was found to be better according to the outcomes of the fuzzy assessment method of TFNs, The control Department demonstrated a better performance according to the mean values. All those data give an indication that the APOS/ACE style was beneficial for the students of the experimental Department. At any case, more experimental research is needed to allow safer conclusions.

Remark: Observe that in paragraph (ii) of the above example the mean values D_1 and D_2 are linear combination of the TFNS A, B, C, D and F of the form $M=k_1A+k_2B+k_3C+k_4D+k_5F$, with k_i non negative rational numbers, i=1, 2, 3, 4, 5. Consequently, if A (a_1, b_1, c_1), B (a_2, b_2, c_2),….,

F(a_5, b_5, c_5) and M(a, b, c), then

$$M=\sum_{i=1}^{5}k_i(a_i,b_i,c_i)=(\sum_{i=1}^{5}k_ia_i,\sum_{i=1}^{5}k_ib_i,\sum_{i=1}^{5}k_ic_i)$$

$$X(M)=\frac{\sum_{i=1}^{5}k_ia_i+\sum_{i=1}^{5}k_ib_i+\sum_{i=1}^{5}k_ic_i}{3}=\sum_{i=1}^{5}k_i\frac{a_i+b_i+c_i}{3}$$

Therefore,

$$=\sum_{i=1}^{5}k_ib_i=b$$.

This shows that in practice, in order to defuzzify D_1 and D_2 it is enough to calculate their *middle entries* only. Obviously this method can be used in all cases, even if the mean values are comparable to each other TFNs.

ASSESSMENT OF STUDENT LEARNING SKILLS ACCORDING TO THE BLOOM'S TAXONOMY

The Bloom's Taxonomy for Teaching and Learning

In 1956 Benjamin Bloom with collaborators Max Englehart, Edward Furst, Walter Hill, and David Krathwohl published a framework for categorizing educational goals, the ***Taxonomy of Educational Objectives*** [3] [1]. Although named after Bloom, the publication of the taxonomy followed a series of conferences from 1949 to 1953, which were designed to improve communication between educators on the design of curricula and examinations. A revised version of the taxonomy was created in 2000 by Lorin Anderson [1], former student of Bloom. Since the taxonomy reflects different forms of thinking and thinking is an active process, in the revised version the names of its six major levels were changed from noun to verb forms. The six major levels of the revised taxonomy are presented in Figure 1, taken from Wikipedia, 2015.

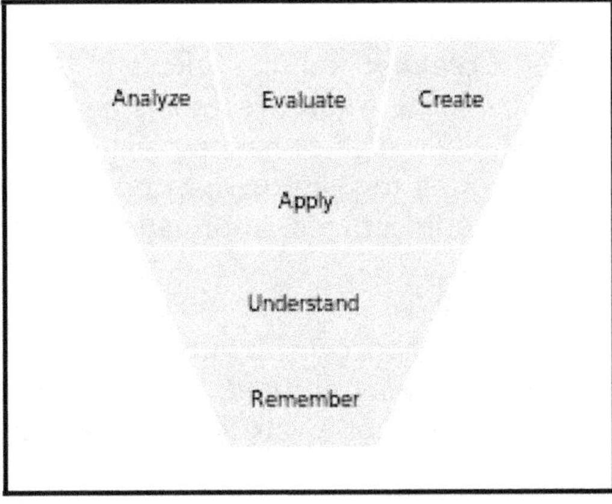

Figure 4: The six major levels of the Bloom's taxonomy

The above six levels in the taxonomy, moving through the lowest order processes to the highest, could be described as follows :

- ***Knowing - Remembering:*** Retrieving, recognizing, and recalling relevant knowledge from long-term memory, e.g. find out, learn terms, facts, methods, procedures, concepts

- ***Organizing - Understanding:*** Constructing meaning from oral, written, and graphic messages through interpreting, exemplifying, classifying, summarizing, inferring, comparing, and explaining. Understand uses and implications of terms, facts, methods, procedures, concepts.

- ***Applying:*** Carrying out or using a procedure through executing, or implementing. Make use of, apply practice theory, solve problems, use information in new situations.

- ***Analyzing:*** Breaking material into constituent parts, determining how the parts relate to one another and to an overall structure or purpose through differentiating, organizing, and attributing. Take concepts apart, break them down, analyze structure, recognize assumptions and poor logic, evaluate relevancy.

- ***Generating - Evaluating:*** Making judgments based on criteria and standards through checking and critiquing. Set standards, judge using standards, evidence, rubrics, accept or reject on basis of criteria.

- ***Integrating - Creating:*** Putting elements together to form a coherent or functional whole; reorganizing elements into a new pattern or structure through generating, planning, or producing. Put things together; bring together various parts; write theme, present speech, plan experiment, put information together in a new & creative way

Most researchers and educators consider the last three levels --analyzing, evaluating and creating – as being *parallel*, i.e. as happening together. It is obvious that using Bloom's higher levels helps the students become better problem solvers.

For teaching a topic, the instructor should arrange his/her class work in the order to synchronize it with these six steps of Bloom's Taxonomy. The typical questions for evaluating the student achievement at the corresponding level are the following:

Knowing questions focus on clarifying, recalling, naming, and listing:

Which illustrates...?

Write... in standard form....

What is the correct way to write the number of... in word form?

Organizing questions focus on arranging information, comparing similarities/ differences, classifying, and sequencing:

Which shows... in order from...?

What is the order...?

Which is the difference between a... and a...?
Which is the same as...?

Express... as a...?

Applying questions focus on prior knowledge to solve a problem:

What was the total...?

What is the value of...?

How many... would be needed for...?

Solve....Add/subtract....Find....Evaluate....Estimate....Graph....

Analyzing questions focus on examining parts, identifying attributes/ relationships/patterns, and main idea:

Which tells...?

If the pattern continues, ...

Which could...?

What rule explains/completes... this pattern?

What is/are missing?

What is the best estimate for...?

Which shows...?

What is the effect of...?

Generating questions focus on producing new information, inferring, predicting, and elaborating with details:

What number does... stand for?

What is the probability...?

What are the chances...?

What effect...?

Integrating questions focus on connecting/combining/summarizing information, and restructuring existing information to incorporate new information:

How many different...?
What happens to... when...?

What is the significance of...?

How many different combinations...?

Find the number of..., ..., and ... in the figure below.

Evaluating questions focus on reasonableness and quality of ideas, criteria for making judgments and confirming accuracy of claims:

Which most accurately...?

Which is correct?

Which statement about... is true?

What are the chances...?

Which would best...?

Which would... the same...?

Which statement is sufficient to proven...?

Bloom's taxonomy serves as the backbone of many teaching philosophies, in particular those that lean more towards skills rather than content. The emphasis on higher-order thinking inherent in such philosophies is based on the top levels of the taxonomy including analysis, evaluation, synthesis and creation. Bloom's taxonomy can be used as a teaching tool to help balance assessment and evaluative questions in class, assignments and texts to ensure all orders of thinking are exercised in student's learning.

A Classroom Application

The following application was performed with subjects the students of two different departments (30 students in each department) of the School of Technological Applications (prospective engineers) of the Graduate Technological Educational Institute (T. E. I.) of Western Greece attending the common course "Mathematics I" of their first term of studies and having the same instructor (Voskoglou, 2016a). This course involves an introductory module repeating and extending the students' knowledge from secondary education about the real numbers. After the module was taught, the instructor wanted to investigate the students' progress according to the principles of the Bloom's Taxonomy. For this, he asked them to answer in the class the written test presented at the end of this paragraph, which is divided in six different parts, one for each level of the Taxonomy. The students' answers were assessed separately for each level in a scale from 0 to 100 and the means obtained correspond to each student's overall performance.

Denote by L_i, i=1, 2, 3, 4, 5, 6 the levels of Knowing-Remembering, Organizing-Understanding, Applying, Analyzing, Generating-Evaluating and Integrating- Creating respectively of the Bloom's Taxonomy and by P the student overall performance. Then the test's results are depicted in the following two tables:

Table 2: Results of the first Department

Grade	L_1	L_2	L_3	L_4	L_5	L_6	P
A(85100)	8	6	5	3	2	3	4
B(84-75)	9	11	10	8	7	8	9
C(74-60)	10	9	10	12	10	8	10
D(59-50)	3	3	3	5	7	8	5
F(<50)	0	1	2	2	4	3	2

Table 3: Results of the second Department

Grade	L_1	L_2	L_3	L_4	L_5	L_6	P
A(85100)	9	8	6	4	3	3	5
B(84-75)	6	7	9	7	7	6	8
C(74-60)	9	8	10	12	10	8	9
D(59-50)	6	7	4	4	7	11	7
F(<50)	0	0	1	3	3	2	1

For reasons of simplicity, let us denote the mean values of each department's performance at level L_i, i =1, 2,…, 6 and its overall performance P by the same letters. Then, from Table 2 one finds that for the first department $L_1 = \frac{1}{30}(8A+9B+10C+3D) = \frac{1}{30}(2105, 2287.5, 2473) \approx (70.17, 76.25, 82.43)$, which gives that $x(L_1) \approx 76.25$.

Working in the same way one also finds that $x(L_2) \approx 74.02$, $x(L_3) \approx 71.33$, $x(L_4) \approx 67.97$, , $x(L_5) \approx 63.33$, $x(L_6) = 65.3$ and $x(P) \approx 69.23$. The above outcomes show that the first department demonstrated a very good (B) performance at level L_1 of the Bloom's Taxonomy, a good (C) performance at all the other levels and a good overall performance as well.

Similarly, from Table 3 and for the second department one finds that $x(L_1) = 74.65$, $x(L_2) = 73.8$, $x(L_3) \approx 72.77$, $x(L_4) = 67.4$, $x(L_5) = 65.3$, $x(L_6) = 61.3$ and $x(P) =70.25$. The above outcomes show that the second department demonstrated a good performance at all levels of the Taxonomy and a good overall performance as well.

On comparing the outcomes of the two departments one concludes that the first department demonstrated a better performance at levels L_1, L_2, L_4 and L_6 of the Bloom's Taxonomy, while the second department demonstrated a better performance at levels L_3 and L_5 and a better overall performance than the first department.

Observe also that the performance of each department is decreasing from level L_1 to level L_4, which was expected, since the success at the higher levels is based on the lower levels. However, for the first department this does not happen for the last three levels, a fact which is compatible to the view of most researchers and educators that the three higher levels of the Taxonomy are parallel to each other.

The questionnaire used in the application

Topic: Real numbers (introductory College level)

1. Knowing - Remembering

- Give the definitions and examples of a periodic decimal and of an irrational number (in the form of an infinite decimal).

2. Organizing

- Compare the set of all fractions with the set of periodic decimals. Compare the set of irrational numbers with the set of all roots (of any order) that have no exact values.

3. Applying

- Which of the following numbers are natural, integers, rational, irrational and real numbers?

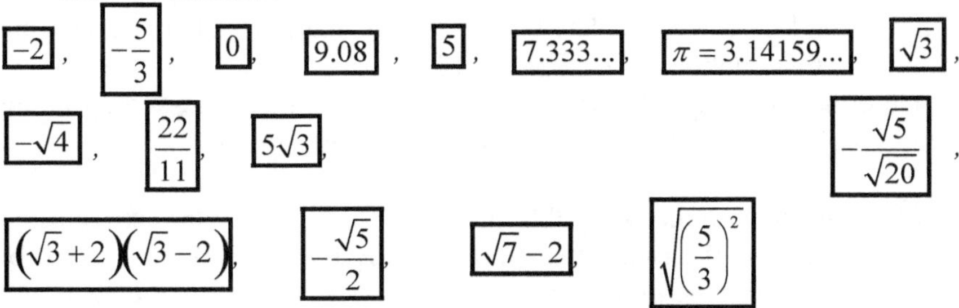

4. Analyzing

- Find the digit which is in the 1005th place of the decimal 2.825342342......
- Write the number 0.345345345... in its fractional form.
- Compare the numbers 5 and 4.9999…
- Construct the line segment of length $\sqrt{3}$ with the help of the Pythagorean Theorem. Give a geometric interpretation.

5. Generating- Evaluating

- Justify why the decimals 2.00131311311131111... and 0.1234567891011... are irrational numbers.
- Construct the line segment of length $\sqrt[3]{2}$ by using the graph of the function f(x)=$\sqrt[3]{x}$

6. Integrating- Creating:

- Define the set of the real numbers in terms of their decimal representations (this definition was not given by the instructor to the class before the test).

BASKET-BALL PLAYER ASSESSMENT

The individual performance of the five players of a basket-ball team who started a game was assessed by six different athletic journalists in a scale from 0 to 100 as follows: **P$_1$** (player 1)**:** 43, 48, 49, 49, 50, 52, **P$_2$:** 81, 83. 85, 88, 91, 95, **P$_3$:** 76, 82, 89, 95, 95, 98, **P$_4$:** 86, 86, 87, 87, 87, 88 and **P$_5$:** 35, 40, 44, 52, 59, 62. The players' performance is characterized by the fuzzy linguistic labels A, B, C, D and F.

Here the five players' overall performance will be assessed by using the traditional methods and also by using TFNs and TpFNs

.

i) Traditional methods: Adding the 5 * 6 = 30 in total scores assigned to the players by the journalists and dividing the corresponding sum by 30 one finds that the *mean performance* of the five players is approximately equal to 60.05, i.e. it can be marginally characterized as good (C).

Further, it can be observed that 14 of the above scores correspond to the label A, four to B, one to C, four to D and seven to F. Replacing these values to formula (1) of Chapter 6 one finds that the GPA index is approximately equal to 2.47. Therefore, the five players' overall *quality performance* is characterized as more than satisfactory.

ii) Using TFNS: We use again the TFNs A, B, C, D and F of the previous section.. It becomes clear that 14 of the student scores correspond to the TFN A, four to B, one to C, four to D and seven to F.

The mean value of the above TFNs (Definition 12) is equal to

M = $\frac{1}{30}$(14A + 4B + C + 4D + 7F) $\approx$(58.33, 68.98, 79.63). Therefore, the players' overall performance lies in the interval [58.33, 79.63], i.e. it could be characterized from fair (D) to very good (B).

Defuzzifying the TFN M (Proposition 7) one finds the value 68.98 for the x-coordinate of the COG of its graph, which shows that the five players demonstrated a good (C) overall performance.

iii) Using TpFNs: A TpFN (denoted, for simplicity, by the same letter) is assigned to each basket-ball player **P$_i$** as follows: **P$_1$** = (0, 43, 52, 59), **P$_2$** = (75, 81, 95, 100), **P$_3$** = (75, 76, 98, 100), **P$_4$** = (85, 86, 88, 100) and **P$_5$** = (0**,** 35, 62, 74). Each of the above TpFNs describes numerically the individual performance of the corresponding player in the form (a_1, a_2, a_3, a_4), where a_1 and a_4 are the lower and upper bounds respectively of his performance with respect to the linguistic labels A, B, C, D and F, while a_2 and a_3 are the lower and higher scores respectively assigned to the corresponding player by the athletic journalists. The mean value of the

TpFNs $\mathbf{P}_i$, i =1, 2, 3, 4, 5 (Definition 12), is equal to $\mathbf{P} = \frac{1}{5}\sum_{i=1}^{5} P_i$ = (47, 64.2, 79, 86.6), which gives the following information:

- The players' performance, is fluctuated from unsatisfactory (a_1=47) to excellent (a_4=86.6).
- The overall players' performance is lying in the interval [a_2, a_3] = [64.2, 79], i.e. it can be characterized from good (C) to very good (B).

Further, defuzzifying the TpFN **P** (Proposition 11) one finds the value $\frac{79^2+(86.6)^2-(64.2)^2-47^2+79*86.6-47*(64.2)}{3(79+86.6-47-64.2)} \approx 68.84$ for the x-coordinate of the COG of its graph, which shows that the five players demonstrated a good (C) overall performance.

OTHER APPLICATIONS

The combined use of TFNs/TpFNs and of the COG technique was also applied in earlier author's works for the assessment of learning skills (Voskoglou, 2015a), of AR skills (Voskoglou & Subbotin, 2015a), of MM skills (Voskoglou & Subbotin, 2016), for the verification of a chosen decision (Voskoglou, 2015b), for the effectiveness of a CBR system (Voskoglou & Subbotin, 2015b), for the Bridge player performance ((Voskoglou 2016c), for the student ability of understanding the infinity (Voskoglou 2016b), etc.

CONCLUSIONS

I) A combination of TFNS (or TpFNs) and of the COG technique was used in this Chapter for assessing student (APOS/ACE, Bloom's Taxonomy) and basket-ball player performance The outcomes obtained in our Examples are depicted in Tables 3 and 4 below.

Table 3: Outcomes of student assessment (APOS/ACE)

Method	D_1	D_2	Performance
Mean values	73.79	72.39	Good (C)
GPA index	2.53	2.53	More than satisfactory
TFNs	71.74	72.71	Good (C)

Table 4: Outcomes of player assessment

Method		Performance
Mean values	60.05	Good (C)
GPA index	2.47	More than satisfactory
TFNs	68.98	Good (C)
TpFNs	68.84	Good (C)

Observing these Tables it turns out that the conclusions obtained by the two new fuzzy assessment methods (TFNs / TpFNs) are compatible to the corresponding conclusions of the two traditional assessment methods (mean values / GPA index). This gives a strong indication that the new fuzzy methods "behave" well (validation of the fuzzy methods), which was also crossed with different problems in earlier author's works (Voskoglou, 2015d, e, Voskoglou, 2016a, b, c, Voskoglou & Subbotin, 2016, etc.). The small numerical differences appeared are due to the different "philosophy" of the four methods (mean and quality performance, bi-valued and fuzzy logic).

II)The approximation of the overall player performance obtained in the first step of case (iii) of the corresponding Example (good to very good) is better than that obtained in the first step of case (ii) (fair to very good), because the TpFNs, due to the way of their construction, assess more accurately than the TFNs the player performance. However, it is not always easy in practice to use TpFNs instead of TFNs. In the APOS/ACE Example, for instance, there are 170+255 = 425 students under assessment, which means that one has to form a great number of TpFNs for evaluating their performance, resulting to laborious calculations.

III) Another advantage of using TpFNs as assessment tools is that they enable, in contrast to TFNs, the comparison of the ***individual performance***

of any two persons. To do this in case (iii) of the Example with the basket-ball players, for instance, it is enough to defuzzify the TpFNs $\mathbf{P}_i$, i=1, 2, 3, 4, 5 corresponding to each player's individual performance, as we did for **P**, and to apply the same comparison criterion.

REFERENCES

Anderson, L. W., Krathwohl, D.R. (2000), *A taxonomy for learning, teaching, and assessing: A revision of Bloom's taxonomy of educational objectives*, Allyn and Bacon, Boston.

Asiala, M., et al. (1996)., A framework for research and curriculum development in undergraduate mathematics education, *Research in Collegiate Mathematics Education II, CBMS Issues in Mathematics Education*, 6, 1-32,.

Bloom, B. S., Engelhart, M. D., Furst, E. J., Hill, W. H., Krathwohl, D. R. (1956), *Taxonomy of educational objectives: The classification of educational goals*, Handbook I: Cognitive domain, David McKay Company, New York.

Confrey, J. (1995), How compatible are radical constructivism, socio-cultural approaches and social constructivism?, in L. P. Steffe & J. Gale (Eds.): *Constructivism in Education*, Lawrence Erlbaum Associates, Hillslade, NJ.

Davis, R. B., Maher, C. A. & Noddings, N. (1990), Constructivist Views on the Learning and Teaching of Mathematics, *Journal for Research in Mathematics Education*, Monograph No. 4, National Council of Teachers of Mathematics, Reston, Virginia.

Dubinsky, E. & McDonald, M. A. (2001), APOS: A constructivist theory of learning in undergraduate mathematics education research. In: D. Holton et al. (Eds.), *The Teaching and learning of Mathematics at University Level: An ICMI Study*, 273-280, Kluwer Academic Publishers, Dordrecht, Netherlands.

Kaufmann, A. & Gupta, M. (1991), *Introduction to Fuzzy Arithmetic*, Van Nostrand Reinhold Company, , New York.

Klir, G.J. & Folger, T.A. (1988), *Fuzzy Sets, Uncertainty and Information*, Prentice-Hall, London.

Maharaj, A. (2013), An APOS analysis of natural science students' understanding of derivatives, *South African Journal of Education,* 33(1), 19 pages.

Piaget, J. (1970), *Genetic Epistemology*, Columbia University Press, New York and London,.

Sakawa, M.(1993), *Fuzzy Sets and Interactive Multiobjective Optimization*, Plenum press, NY and London.

Voskoglou, M.Gr. (2013), An application of the APOS/ACE approach in teaching the irrational numbers, Journal of Mathematical Sciences and Mathematics Education, 8(1), 30-47.

Voskoglou, M. Gr. (2015a), An Application of Triangular Fuzzy Numbers to Learning Assessment, *Journal of Physical Sciences,* 20, 63-79.

Voskoglou, M. Gr. (2015b), Using the Triangular Fuzzy Numbers for the Verification of a Chosen Decision, *Academic Journal of Applied Mathematical Sciences,* 1(1), 1-8

Voskoglou, M.Gr. & Subbotin, I.Ya. (2015a), Application of the Triangular Fuzzy Assessment Model to Assessment of Analogical Reasoning Skills, *American Journal of Applied Mathematics and Statistics,* 3(1), 1-6.

Voskoglou, M.Gr. & Subbotin, I.Ya. (2015b), Evaluating the Effectiveness of a CBR System, *American Journal of Computational and Applied Mathematics,* 5(2), 27-32.

Voskoglou, M.Gr. (2016a), Fuzzy numbers as an assessment tool in the APOS/ACE instructional treatment of mathematics, *Journal of Physical and Mathematical Education*, Sumy State Pedagogical University, Ukraine, 1(7), 29-37.

Voskoglou, M.Gr. (2016b), A tool for assessing the ability of understanding the infinity based on triangular fuzzy numbers, *Egyptian Computer Science Journal*, 40(2), 11-23.

Voskoglou, M.Gr. (2016c), An application of fuzzy numbers to assessment of bridge players performance, *American Journal of Business and Society*, 1(2), 32-41.

Voskoglou, M.Gr. & Subbotin, I.Ya. (2016), An application of fuzzy numbers to the assessment of mathematical modelling skills, *International Journal of Mathematical Modelling and Computations*, 6(1), 83-103.

Weller, K. et al. (2003), Students performance and attitudes in courses based on APOS theory and the ACE teaching cycle. In: A. Selden et al. (Eds.), *Research in collegiate mathematics education,* V, pp. 97-181, Providence, RI: American Mathematical Society.

Weller, K. , Arnon, I & Dubinski, E. (2009) , Pre-service Teachers' Understanding of the Relation Between a Fraction or Integer and Its Decimal Expansion, *Canadian Journal of Science, Mathematics and Technology Education,* 9(1), 5-28.

Weller, K. , Arnon, I & Dubinski, E. (2011), Pre-service Teachers' Understanding of the Relation Between a Fraction or Integer and Its Decimal Expansion: Strength and Stability of Belief, *Canadian Journal of Science, Mathematics and Technology Education,* 11(2), 129-159.

Wikipedia (2015), Bloom's Taxonomy, retrieved on February 10 from: http://en.wikipedia.org/wiki/Bloom's_taxonomy .

ENDNOTE

Bloom's taxonomy divides educational objectives into three domains: ***cognitive***, ***affective*** and ***psychomotor***, sometimes loosely described as "knowing/head", "feeling/heart" and "doing/hands" respectively. The volume published in 1956 [5] and the revision followed in 2000 [6] concern the cognitive domain, while a second volume published in 1965 on the affective domain. A third volume was planned on the psychomotor domain, but it was never published. However, other authors published their own taxonomies on the last domain. More details can be found in Wikipedia, 2015.

CHAPTER 8
Conclusions and Perspectives for Future Research

ABSTRACT
The Chapter starts with a recapitulating application to Computational Thinking of the methods and techniques used earlier in this book. The general conclusions drawn from the discussion performed in the book follow and the Chapter closes with some hints for future research on the topics discussed.

A RECAPITULATING APPLICATION

Computational Thinking (CT)

As we have seen in Chapter 6 the complexity of the day to day life in our modern society has made ***Critical thinking (CrT)*** a necessary condition for the solution of non routine problems. Nevertheless, CrT is not also a sufficient condition for PS, especially when tackling complicated technological problems, where computers are frequently used as a supporting tool. In this case the need for ***Computational Thinking (CT)*** becomes another prerequisite for PS. Computation is nowadays an increasingly essential tool for doing scientific research. It is expected that future generations of scientists and engineers will need to engage and understand computing in order to work effectively with information systems, technologies and methodologies. CT, named so for its extensive use of computer science techniques (Wing, 2006), is a type of analytical thinking that employs mathematical and engineering thinking to understand and solve complex problems within the constraints of the real world.

The relationship between CT and CrT, the two basic modes of thinking for PS, has not been clearly established yet. In Voskoglou & Buckley, 2012, we have attempted to shed some light into this relationship. The conclusions of our study can be summarized with the help of Figure 1, where a 3 - dimensional model for the PS process is presented. According to this model the existing ***knowledge*** serves as the connecting tool between CrT and CT, while the problem's solution appears to be the "product" of a simultaneous application of the above three components (knowledge, CrT and CT) to the PS process. This approach is based on the hypothesis that, when the already existing knowledge is adequate, the necessary for the problem's solution new knowledge is obtained through CrT, while CT is applied to design and construct the solution.

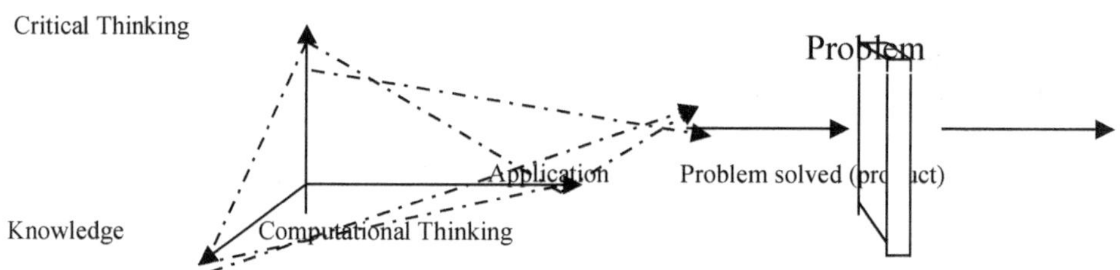

Figure 1: The 3- dimensional model for the PS process

The type of each problem dictates the order of the application of the above three components, which (order) can have in certain, relatively simple, cases the linear form of Figure 2.

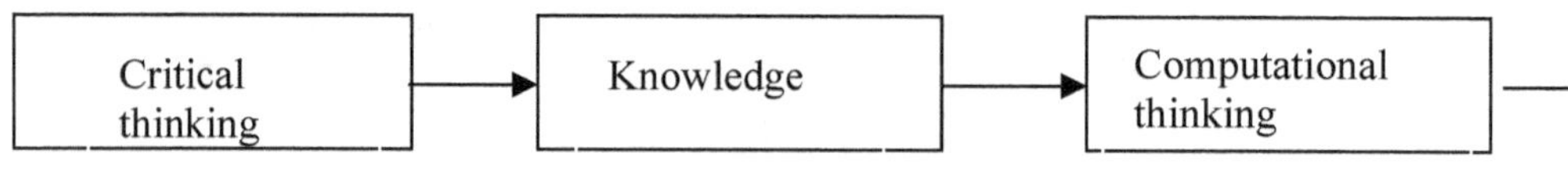

Figure 2: The linear PS model

The above model (Figures 1 and 2) can be used in formulating the PS process of the complex problems of our everyday life and especially of the complicated technological problems. According to Liu and Wang [7] CT is a hybrid of other modes of thinking including ***abstract, logical, modelling and constructive thinking***. It synthesizes the existing knowledge with CrT and applies them for modelling and solving complicated problems, for building engineering systems, for interpreting data, etc

Polar Coordinates in the Plane

A ***coordinate system*** can be defined in general as a rule for mapping pairs of numbers to points of the plane (or the space).There are several coordinate systems used in mathematics, engineering and the other applied sciences, all of them based on this idea.

Descartes (1596-1650) introduced in 1637 the (x, y)-coordinates, known nowadays as the ***Cartesian coordinates***, which opened the door to the development of Analytic Geometry.

Another popular coordinate system with many applications to differential and integral calculus, to complex numbers, to other branches of pure and applied mathematics, to physics, etc., is the system of ***Polar Coordinates (PCO)***. It is recalled that the PCO of a point P of the plane are defined by a pair of numbers (r, θ), where r is the algebraic distance of P from a fixed point O of the plane, called the origin, and θ is the angle formed by the polar semi-axis OX and the straight line segment OP. The numbers r and θ can be positive, negative or zero (Figure 3).

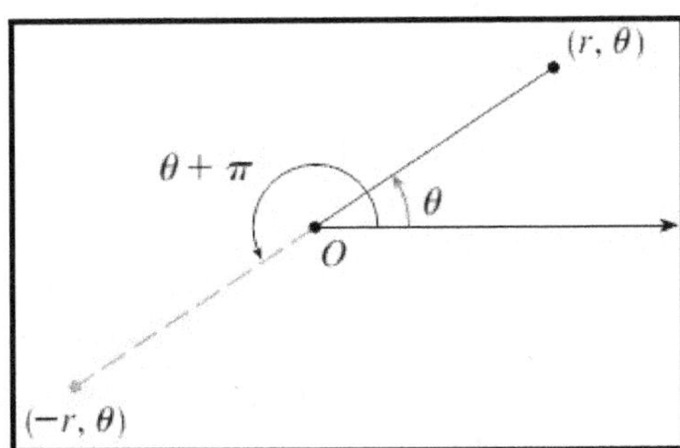

Figure3: Polar coordinates of a point in the plane

It becomes evident that, in contrast to its Cartesian coordinates, each point P of the plane has not a unique polar representation. PCO can be also defined in the 3-dimensional space, but this is out of the scope of the present application.

In recent works (Borji & Voskoglou, 2016, 2017) we have studied the student difficulties for understanding and properly use the PCO in the plane and he have designed an ACE circle (see Chapter7) for teaching and learning them in introductory university level. A classroom experiment (Borji & Voskoglou, 2017) will be described here connecting the student understanding of PCO with the use of computers.

The Classroom Experiment

Two groups of university students participated in this research. The first group (***control group***) was enrolled in a calculus course in the fall semester of 2016 including the study of the PCO in the plane. Eighteen students participated in this class and the topic of the PCO was taught by the instructor in the traditional, lecture - based method.

The second group (***experimental group***) was enrolled in a calculus course in another university at the same semester. Twenty students participated in this class, in which the topic of the PCO in the plane was taught on the basis of the ACE circle designed by Borji and Voskoglou (2017). The ACE circle involved the following three iterations:

1. Prerequisite knowledge (Cartesian coordinates, generalized angles, trigonometric functions).

2. Identifying and plotting points of the plane by their PCO and converting Cartesian to PCO and vice versa.

3. Sketching graphs of polar equations in all quadrants of the plane.

The ***Maple*** software was used in those iterations for facilitating the student understanding. This means that approximately half of the time available for the lectures, which was the same for the two groups, was spent in a computer laboratory for the experimental group. Consequently, since working with computers consumes more time, the students of the control group were exposed to more examples on the board than those performed for the students of the experimental group.

Before starting the teaching of PCO both groups completed a written ***pre-test***. Since angles, trigonometric functions and Cartesian equations are fundamental prerequisites for learning polar coordinates and, as we found in our relevant research (Borji & Voskoglou, 2016), students face usually problems when dealing with them, the pre-test's questions, were based on those concepts. The pre-test's results demonstrated approximately the same performance for the two groups.

The Post-Test

One week after the end of lectures on PCO a post-test was performed for both groups. For reasons of justice the questions of the post-test were designed by other mathematics faculties who are familiar with polar coordinates and none of the two instructors participated in their design. The questions of the post-test were the following:

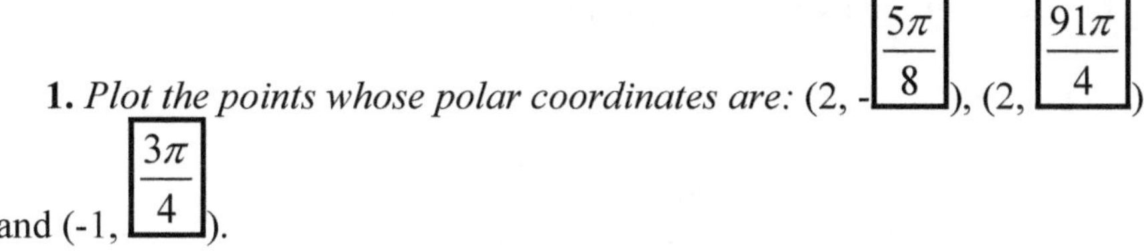

1. *Plot the points whose polar coordinates are:* $(2, -\frac{5\pi}{8})$, $(2, \frac{91\pi}{4})$ and $(-1, \frac{3\pi}{4})$.

2. *Find the Cartesian coordinates of the point* $(3, \frac{9\pi}{4})$.

3. *Find the polar coordinates of the point*

4. *Find the polar equation for the curve*

5. *Identify the curve by finding the Cartesian equation for it.*

6. *Sketch the curve with polar equation*

7. *The figure below shows a graph of r as a function of* θ *in Cartesian coordinates. Use it to sketch the corresponding polar curve.*

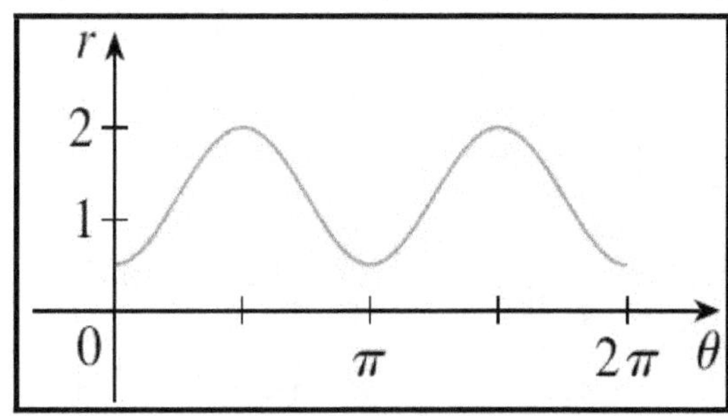

Figure 4: The Cartesian graph of r = f(θ)

The student papers were marked together by both instructors for the control and the experimental group with scores ranging from 0–100. The student scores were the following:

Control group (G_1): 88, 75, 62, 55, 55, 45, 40, 36, 32, 30, 27, 27, 25, 20, 18, 15, 15, 10.

Experimental group (G_2): 100, 85, 80, 76, 76 65, 60, 60, 55, 48, 45, 42, 35, 30, 28, 25, 20, 20, 15, 15.

The results of the two groups are depicted in Table 1.

Table 1: Post-Test Results

Grades	G_1	G_2
A	1	2
B	1	3
C	1	3
D	2	1
F	13	11
Total	18	20

Recapitulating the contents of Part II of this book, the group uncertainty will be measured and their performance will be evaluated with all the traditional and fuzzy assessment methods used in the book:

(i) Group Uncertainty: From Table 1 one finds for G_1 that $m(\mathrm{A}) = m(\mathrm{B}) = m(\mathrm{C}) = \frac{1}{18}$, $m(\mathrm{D}) = \frac{2}{18}$ and $m(\mathrm{F}) = \frac{13}{18}$. Therefore the ordered possibility distribution is

r: $r_1 = 1 > r_2 = \frac{2}{13} > r_3 = r_4 = r_5 = \frac{1}{13}$ and equation (2) of Chapter 5 gives that the strife $ST(r) = \frac{1}{13 \log 2} \log\left(\frac{26}{15}\right) \approx 0.06$. Also, equation (3) of

Chapter 5 gives that the non-specificity $N(r) = \frac{1}{13} \approx 0.08$. Therefore, the total possibilistic uncertainty for G_1 is equal to $T(r) \approx 0.14$.

Similarly, for G_2 one finds from Table 1 that $m(A) = \frac{2}{20}$ $m(B) = m(C) = \frac{3}{20}$,

$m(D) = \frac{1}{20}$ and $m(F) = \frac{11}{20}$. Therefore, $\boldsymbol{r}$: $r_1 = 1 > r_2 = r_3 = \frac{3}{11} > r_4 = \frac{2}{11} > r_5 = \frac{1}{11}$,

$ST(r) = \frac{1}{log2}\left[\frac{1}{11} \log\left(\frac{33}{17}\right) + \frac{1}{11} \log\left(\frac{44}{19}\right)\right] \approx 0.2$, $N(r) = \frac{1}{11 log2}(\log 3 + 2\log 2) \approx 0.33$ and $T(r) \approx 0.53$

On the other hand, replacing the membership degrees to equation (1) of Chapter 5 one finds that the ***probabilistic uncertainty*** is

$$H= -\frac{1}{\ln 5}\left(\frac{13}{18}\ln\frac{13}{18} + \frac{2}{18}\ln\frac{2}{18} + \frac{3}{18}\ln\frac{1}{18}\right) \approx 0.59 \text{ for } G_1 \text{ and}$$

$$H= -\frac{1}{\ln 5}\left(\frac{11}{20}\ln\frac{11}{20} + \frac{6}{20}\ln\frac{3}{20} + \frac{2}{20}\ln\frac{2}{20} + \frac{1}{20}\ln\frac{1}{20}\right) \approx 0.8 \text{ for } G_2.$$

(ii) ***Mean values:*** It is straightforward to check that the mean values of the student scores are 37.5 and 49 for the control and the experimental group respectively demonstrating a better performance for G_2, but a non satisfactory mean performance for both groups.

(iii) GPA index: Replacing the data of Table 1 to equation (1) of Chapter 6 one finds that GPA$=\frac{4+3+2+2}{18} \approx 0.61$ for G_1 and GPA$=\frac{8+9+6+1}{20} = 1.2$ for G_2. These values, being smaller than the half of the GPA's maximal value which is 2, demonstrate a less than satisfactory quality performance for both groups, but a better performance for the experimental group G_2.

iv) RFAM: By Corollary 4 of Chapter 6, the experimental group demonstrates a better quality performance. However, replacing the values of the GPA index to formula (13) of Chapter 6 one finds for RFAM that $x_c = 0.5 + 0.61 = 1.11$ for G_1 and $x_c = 0.5 + 1.2 = 1.7$ for G_2. Since the above values are less than 2.25, according to RFAM both groups demonstrate a non satisfactory quality performance.

v) GRFAM, TFAM and ***TpFAM:*** By Corollary 4 of Chapter 6, the experimental group demonstrates a better quality performance. However, replacing the values of the GPA index to formula (12) of Chapter 6 one finds for the above equivalent variations of RFAM that $X_c = 0.5 + 0.7*0.61 = 0.93$ for G_1 and $X_c = 0.5 + 0.7*1.2 = 1.34$ for G_2. Therefore, since those values are less than 1.65, according to GRFAM, TFAM and TpFAM both groups demonstrate a non satisfactory quality performance.

(vi)TFNs: We consider the TFNs A, B, C, D and F, as we have defined them in Chapter 7. The mean values of those TFNs are equal to

$G_1 = \frac{1}{18}(A + B + C + 2D + 13F) = (1013, 666.5, 320) \approx (56.28, 37.03, 17.78)$, with $X(G_1) \approx 37.03$ for the control group and

$G_2 = \frac{1}{20}(2A + 3B + 3C + D + 11F) = (1272, 948.5, 625) \approx (63.6, 47.43, 31.25)$, with $X(G_1) \approx 47.43$ for the experimental group.

Therefore, both groups demonstrated a non satisfactory mean performance, with the performance of the experimental group being better. Decision – Making

In concluding, although both groups demonstrated a non satisfactory mean and quality performance, the use of computers by the students of the experimental group helped them for a better understanding of the POC with respect to the control group.

GENERAL CONCLUSIONS

Elements from the theory of ***finite Markov Chains*** and principles of ***Fuzzy Logic*** were used in this book for the description and assessment of several learning contexts, and not only, like Problem - Solving, Mathematical Modelling, Learning a subject matter in the classroom, Iterative Learning, Learning on the basis of the Bloom's taxonomy, Teaching and Learning mathematics with the APOS/ACE instructional treatment, Critical and Computational Thinking, Analogical and Case-

Based Reasoning, Van-Hiele levels of Geometric Reasoning, student understanding of the Infinity, Decision – Making, Market's Research and other applications to Management, assessment of basket-ball and bridge player performance, etc.

The general Markov Chain model gave us the possibility to make ***short-run*** forecasts for the evolution of the corresponding situations, while the theory of ***Ergodic Chains*** enabled us to make ***long-run*** forecasts where the existing conditions are stabilized (equilibrium situation). Finally, using principles of the theory of ***Absorbing Chains*** we obtained measures for the effectiveness of the corresponding systems and the probabilities of certain processes to terminate in one or another way. However, the Markov chain models, due to the relevant limitations of the Probability Theory on which they are based, are self-restricted to describe only the ***assumed system's behaviour*** during an activity, i.e. the way in which it should act and not how it really acts.

On the contrary, Fuzzy Logic which is based on Zadeh's ***fuzzy sets*** has no such limitations. Due to its property of characterizing the ambiguous cases with multiple values it opens the door to the solution of real life problems expressed in a natural language, it is able to describe the ***acting to the moment*** human behavior and provides rich resources for assessment under conditions characterized by uncertainty, vagueness or ambiguity. In this book we developed a general fuzzy model for the description of a fuzzy system's activities, which is based on the calculation of the ***possibilities*** of all profiles of the system's elements participating in the corresponding activity. Several methods were developed as assessment tools of a fuzzy system's performance, including the ***COG defuzzification technique*** (or ***RFAM***) and its variations ***GRFAM, TFAM*** and ***TpFAM***, as well as the use of ***Triangular*** and ***Trapezoidal Fuzzy Numbers.*** All those fuzzy assessment methods were validated by comparing their outcomes with the corresponding outcomes of two traditional assessment methods of the bi-valued logic, the calculation of the ***mean value*** and of the ***GPA index*** of the corresponding scores. Note that the mean value and the method of using the Fuzzy Numbers measure a system's ***mean performance***, while the GPA index, the COG technique and its variations are connected to its ***quality performance*** by assigning greater coefficients to the higher scores.

A system's uncertainty characterizes its capacity to obtain information. In this book we studied two types of a fuzzy system's uncertainty: The ***probabilistic uncertainty,*** which is a generalization of the classical

Shannon's entropy in a fuzzy environment and the ***total possibilistic uncertainty***, which for many researchers is more suitable for the study of the human cognition.

PERSPECTIVES FOR FUTURE RESEARCH

The general character of the Markov Chain and Fuzzy Logic models developed in this work make possible their application to many other sectors of human and machine activities, like Industry, Commerce, Machine Learning and Artificial Intelligence, Sociology, Economics, etc. This is our first target for future research on the subject. Many other scientists have also used, before or after us, similar tools for the same or analogous purposes.

On the other hand there exists a variety of other known tools, which have not been used in the present research, like Markov Chains with fuzzy states, fuzzy matrices, fuzzy graphs, fuzzy relation equations, other defuzzification methods, etc. Such kind of tools could be also used for the development of our models in our future applications.

Also, it is always open the possibility of the development of new mathematical tools to use for the purposes of the relevant research, or the use of already known topics of theoretical mathematics, which have not found any practical applications yet. The latter, although it looks amazing, it commonly happens in the history of Sciences. Characteristic examples are the use by Einstein of the ***Riemann's non-Euclidean Geometry*** for the development of the ***General Relativity Theory,*** the theory of ***Knots***, introduced as an unsuccessful attempt to model the structure of the atom, but finally realized to be the basic tool for studying the DNA mechanisms, and many others.

I want also to mention another example connected to my personal research activities, namely the case of ***Iterative Skew Polynomial Rings***, my Ph.D. Thesis's, purely theoretical on that time, topic (Voskoglou, 1982). In fact, to my great surprise those rings have found recently important applications to the ***Coding Theory*** in Cryptography (Lopez-Permouth, 2009, etc.) and to the Theory of ***Quantum Groups*** (Majid, 2006), which are widely used nowadays for a more effective study of Theoretical Physics.

Examples like the above give strong support to the famous ancient Greek philosopher Plato's theory about the parallel to the real world existence of the "***Universe of Mathematics***". This abstract universe contains all mathematical ideas, methods and principles, which are either known or not known yet to humans. As a result of Plato's theory, which is usually referred as ***Platonism*** and has still many supporters - but and opponents as well - in our era (Shapiro, 2000), humans do not invent the several mathematical topics, but they simply discover them.

REFERENCES

Borji, V. & Voskoglou, M.Gr. (2016), Applying the APOS Theory to Study the Student Understanding of the Polar Coordinates, *American Journal of Educational Research*, 4(16), 1149-1156.

Borji, V. & Voskoglou, M.Gr. (2017), Designing an ACE Approach for Teaching the Polar Coordinates, *American Journal of Educational Research*, 5(3), 303-309..

Liu, J. & Wang, L. (2010), Computational Thinking in Discrete Mathematics, *Proceedings IEEE 2nd International Workshop on Education Technology and Computer Science*, pp. 413-416.

Lopez-Permouth, S. (2009), Matrix Representations of Skew Polynomial Rings with Semisimple Coefficient Rings, *Contemporary Mathematics*, 480, 289-295.

Majid, S. (2006), What is a Quantum Group? *Notices of the American Math. Soc.,* 53, 30-31.

Shapiro, S. (2000), *Thinking about Mathematics,* Oxford University Press, Oxford.

Voskoglou, M.Gr. (1982), *A Contribution to the Study of Rings*, Ph. D. Thesis, Department of Mathematics, University of Patras, Greece (in Greek language).

Voskoglou, M.Gr. & Buckley, S. (2012), Problem Solving and Computers in a Learning Environment, *Egyptian Computer Science Journal*, 36(4), 28-46.

Wing, J.M. (2006), Computational thinking, *Communications of the ACM*, 49, 33-35.

ABOUT THE AUTHOR

Michael Gr. Voskoglou (B.Sc., M.Sc., M. Phil., Ph.D. in Mathematics) is currently an Emeritus Professor of Mathematical Sciences at the School of Technological Applications of the Graduate Technological Educational Institute (T. E. I.) of Western Greece, where he served as a full Professor from1987 to 2010. He used also to be an instructor at the Hellenic Open University, at the Mathematics Department of the University of Patras, at the Schools of Primary and Secondary In – Service Teachers' Training in Patras and a teacher of mathematics at the Greek Public Secondary Education (1972-1987).

As a Visiting Professor he has lectured in M.Sc. courses of the School of Management at the University of Warsaw (2009), of the Department of Operational Mathematics at the University of Applied Sciences in Berlin (2010) and at the Mathematics Department of the Durgapur Institute of Technology under a grand of the GIAN program of the Indian Government. He worked also as a Visiting Researcher at the Institute of Mathematics and Informatics of the Bulgarian Academy of Sciences in Sofia for three years (1997-2000), under sabbatical.

He is the author of 14 books in Greek and in English language and of about 400 papers published in reputed journals and proceedings of conferences of 26 countries of the five continents around the World, with many citations by other researchers. He is the Editor in Chief of the "International Journal of Applications of Fuzzy Sets and Artificial Intelligence" (ISSN 2241-1240), Reviewer of the American Mathematical Society and member of the Editorial Board or referee in many international mathematical journals.

He has conducted five programs of technological research on applications of quantitative methods to Management (1989 – 1997) and he has supervised several students' dissertations of the Graduate T. E. I. of Western Greece on mathematical applications to Economics and Engineering, in the cities of Patras and Mesolonghi. He used also to be an external examiner of Ph.D. dissertations in several universities of Egypt, Saudi Arabia and India.

M. Voskoglou: Finite Markov Chain and Fuzzy Logic Assessment Models

His research interests include Algebra, Fuzzy Logic, Markov Chains, Artificial Intelligence and Mathematics Education.

Links

Home page: http://eclass.teipat.gr/eclass/courses/523102
Related pages: http://arxiv.org/a/voskoglou_m_1
https://mindreaderpublications.academia.edu/MichaelVoskoglou
https://researchgate.net/profile/Michael_Voskoglou
http://www.researcherid.com/rid/C-4504-2014 ; http://orcid.org/0000-0002-4727-0089
https://www.amazon.com/author/michaelvoskoglou
Editor in Chief of IJAFSAI : http://eclass.teipat.gr/eclass/courses/523103
(Impact Factor: SJIF 3.806, UIF 28.4587)

www.ingramcontent.com/pod-product-compliance
Lightning Source LLC
Chambersburg PA
CBHW071420170526
45165CB00001B/344

* 9 7 8 1 5 4 8 3 4 0 0 7 0 *